Rat Jugular Vein and Carotid Artery Catheterization for Acute Survival Studies

Rat Jugular Vein and Carotid Artery Catheterization for Acute Survival Studies

A Practical Guide

Angela Heiser

Foreword by John H.K. Liu

Springer

Angela Heiser
West Roxbury, MA 02132
USA

Library of Congress Control Number: 2006936342

ISBN-10: 0-387-49414-6   E-ISBN-10: 0-387-49416-2
ISBN-13: 978-0-387-49414-2   E-ISBN-13: 978-0-387-49416-6

Printed on acid-free paper.

9 8 7 6 5 4 3 2 1

SPRINGER.COM

CATHETERIZATION

An implant procedure in which a small tube is inserted into a body cavity, duct, or vessel for the purpose of fluid administration or withdrawal ~ *AALAS Reference Directory*

FOREWORD

In biomedical teaching and research, catheterizations of the jugular vein and the carotid artery have been used to access the cardiovascular system in various animal species. Such surgical procedures are technically challenging in rodents because of the small scale involved. There are not many alternatives if the rodent's survival and easy post-operative handling are required. While rodents are becoming predominant research animals, reliable execution of these procedures is essential for many endeavors in pre-clinical research and product development. This practical guide should help scientists and technicians master the procedures in rats and thus confidently move forward to data collection.

For those who have been fortunate to work with Angela Heiser in the laboratory, we have long witnessed how dedicated to the profession a researcher can be. Angie enjoys her work immensely. Her laboratory records are always focused, highly organized, and simply deduced but with in-depth knowledge. In addition to the sciences, she can add stunning, artistic impression to the content. No wonder I have seen so many superb presentations by Angie. Publication of this practical guide certainly will be a milestone.

John H.K. Liu, Ph.D.
Director, Molecular Pharmacology
Hamilton Glaucoma Center
University of California, San Diego
La Jolla, California

ACKNOWLEDGMENTS

 Thanks to my mentors, Drs. John H.K. Liu & Arthur Lage, without whom I would not have been inspired or prepared for this adventure. Thanks to my husband and children for their patience. Thanks to Adria, Jane, Caroline, and Ami for constructive reviews. Thanks to many supporters who encouraged my confidence in this project and the field of animal science, especially Sally Ann. Thanks to Dave at Leica for the generous loan of camera equipment and to Nancy at Scion Pharmaceuticals for the use of their laboratory space. Thanks especially to Jenn for lending a hand when I needed one and to Dr. Lage for his unwavering and continued support.

TABLE OF CONTENTS

INTRODUCTION

Jugular vein and carotid artery catheterizations are among the most widely used surgeries in research labs around the world. Typically, technicians teach these skills to incoming employees and move on to graduate studies or other fields. A resulting "oral history" of catheterization has been passed down from technician to technician. I have endeavored to capture these techniques on paper. Through the years, I have searched for and collected a number of training materials and guides for teaching purposes. This is my attempt to compile all the necessary information in one source.

These catheterizations are extremely important for confirmed intravenous delivery of test substances and arterial blood collection. Catheterization reduces the stress of multiple sampling as observed in association with tail vein or orbital sinus techniques (Ling, 2003; Flynn, 1988; Cocchetto, 1983). Very few adverse effects, including a possible rise of corticosterones and a decrease in platelets, are associated with indwelling catheters except under chronic conditions (Fagin, 1983; Richman, 1980). With practice and dedication to research and humane animal use, one may develop a high throughput paradigm generating predictable results with what is generally thought to be a laborious, rate-limiting step in research. Counting thousands of catheterizations performed in my career, my survival and patency rate are 97%.

> *"We refer to the art of surgery, so why not make it an art and, like the artist, be engrossed in its handicraft."*
> ~ **S. Bunnell, Surgery of the Hand, 1944**.

Some artists have argued that there is an element of science to their approach. I have discovered that surgery also involves an element of art. In fact, in 2004, UCLA organized an exhibit presenting "the world of nanoscience through a participatory aesthetic experience" (nano.arts.ucla.edu). From this basic guide, surgeons will evolve their own subtleties and personal preferences. My sincere hope is that this guide will secure knowledge for future generations of researchers in an original, undiluted format. Society demands a lot from science and this is my contribution.

~ Angie Heiser

PERIOPERATIVE CARE

Perioperative care includes pre-operative, intra-operative, and post-operative care of the patient. Careful attention to these areas is crucial to successful surgery. The goal of this section is to illuminate and define these often-overlooked components of the surgical process.

Pre-operative Care *(prior to surgery)*

- <u>Quarantine</u>: Acclimate incoming rats to their new environment and housing for a minimum of 24 hours. It is preferable to quarantine for 7 – 14 days. This "grace period" allows the animal to return to normal hormonal and metabolic parameters as evidenced by stable body weight following the stress of transport. The investigator can determine that the animals are free of latent or enzootic diseases. Regular handling promotes less stress to the animal and leads to a quicker recovery.
- <u>Pre-operative body weight</u>: Weigh pre-surgery since rats are likely to experience weight loss post-surgery.
- <u>Gross physical exam</u>: Observe the animal for nose or eye discharge, diarrhea, fur matting, and overall appearance and behavior. Palpate for any growths, tumors, or skin abnormalities.
- <u>Antibiotic</u>: If antibiotics are preferred, it is most effective to administer prior to surgery in order to maximize blood levels during surgery and recovery.
- <u>Pre-analgesic</u>: Benefits of analgesics administered pre-surgery are many fold. Rats are less likely to exhibit depression in food intake or experience pain post-surgery and require lower injectable anesthetic doses during surgery. These combined factors contribute to a

quicker recovery and fewer side effects. Single doses are recommended of one of the following; (1) IM or SC flunixin (flunixamine, banamine) @ 1.1 – 3.3 mg/Kg (Stewart, 2003), (2) SC or IV buprenorphine @ 0.01 – 0.05 mg/Kg (Hayes, 1998; Colletti, personal experience).

- <u>Fasting:</u> Do not withhold food prior to surgery unless absolutely necessary because the rat is not likely to consume much post-surgery. An empty stomach prior to surgery is not required since rats are a non-vomiting species (Takeda, 1993).

Intra-operative Care *(during surgery)*

- <u>Mucous membranes</u>: Check the color of the eyes and tongue as an oxygen indicator. A blue color indicates possible hypoxia.
- <u>Body temperature</u>: Body temperature can be monitored. It is important to maintain body heat using a heating pad, heat lamp, or isothermic pad.
- <u>Total blood volume</u>: Avoid hypovolemic shock by controlling blood loss through good hemostasis during surgery.
- <u>Body position</u>: Use care to position the animal in the best possible way to protect cardiac and respiratory function.
- <u>Respiration rate</u>: Watch the animal's breathing by observing the rise and fall of the chest. Observe the pattern and depth of breathing. During surgery, it is possible to monitor by noting the pulse in the carotid artery and the color of the blood in the artery through the scope.
- <u>Tissue handling</u>: Take care to handle tissues gently causing as little disruption as possible. Using a retractor instead of clamped hemostats reduces trauma to skin.

- Wound closure: The use of wound clips or subcuticular suturing may reduce self-mutilation during recovery.

Post-operative Care *(after surgery)*
- Atropine: Atropine sulphate (parasympatholytic) is often used to counteract decreased heart rate due to increased vagal tone. In addition, atropine is used to decrease salivation; allowing the airway to stay open. The surgeon may prefer to administer atropine immediately <u>after</u> surgery to avoid bleeding complications from atropine's effect of increased heart rate. The accepted dose is 0.05 mg/ kg IP. Check expiration date and make fresh.
- Anesthetic recovery: Before the animal is fully awake, 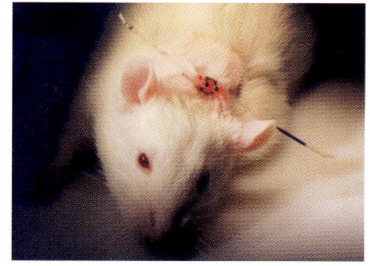 emergencies can occur quickly and unexpectedly. This period of recovery requires the most frequent observation. The animal should be rotated every 30 minutes to avoid edema and irregular breathing. Keep the animal warm. Be careful to avoid heating pad burns from a high heat setting. Regulated heat sources are commercially available such as the ThermoCare ICU unit. Check the wound site for any bleeding.
- Acute recovery: This is a period during which the animal has resumed food and water intake and is approaching normal physiological parameters. The animal should have easy access to food and water. Transgel® or HydroGel™ are effective water sources and can be left on the floor of the homecage with rat chow or Nutra-Gel (food & water source).
- Individual housing: Rats should be housed one to each cage post-surgery. Catheters are better protected from cage mate curiosity and chewing. Some facilities may require this

justification for individual housing because it is normally a more stressful housing condition. Rats prefer group housing (Gentsch, 1982).

- <u>Long-term recovery</u>: During this last stage of recovery, the animal is exhibiting normal physical and behavioral parameters. Rats may be more susceptible to corneal injury following injectable anesthetics (Turner, 2005). Observe the wound site for bleeding, infection, edema, unraveled sutures, and self-mutilation. Sutures can be removed at 7 – 10 days post-surgery. Observations should include motor function and the quantity/quality of urine/feces. Monitor weight gain/loss.

Signs of Pain

Clinical observations of pain and stress in rats can include:
- guarding the wound site
- licking/biting/scratching wound
- vocalization
- rough hair coat/ decreased grooming
- red staining around eyes and nose (porphyrin)
- self-mutilation of wound
- immobility
- decrease in food/water intake
- decreased exploring and grooming
- restlessness
- reluctance to move
- increased respiration
- hunched posture

Ketamine-Xylazine-Acepromazine Cocktail (Adapted from Williams, 1994)

<u>Injection volume:</u>
Female Sprague Dawley:
Intramuscular (IM) 2.2 ml/kg volume, half injected to each thigh

Male Sprague Dawley:
Intramuscular (IM) 2.4 ml/kg volume, half injected to each thigh

<u>Final concentrations:</u>
Ketamine   32.4 mg/ml (71.3 mg/kg)
Xylazine   2.4 mg/ml (5.2 mg/kg)
Acepromazine   0.42 mg/ml (0.9 mg/kg)

<u>Recommended dose ranges (Baker, 1979):</u>
Ketamine   50 – 100 mg/kg IM, IP
Xylazine   1 – 5 mg/kg IM, IP
Acepromazine   2.5 mg/kg IM, IP

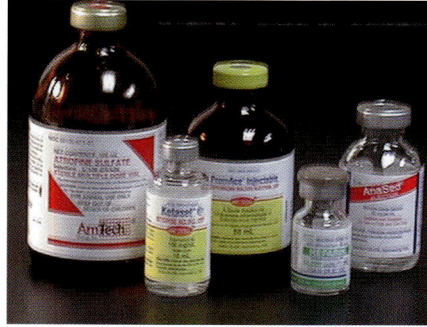

This cocktail can be pre-mixed and stored from two weeks (Johns Hopkins University animal resources) to 6 months (Yale University animal resources) if prepared using sterile methods.

Ketamine Cocktail Dosage Chart *(Sprague Dawley Rats)*

Body Weight (g)	Male (ml)	Female (ml)
230	0.550	0.510
235	0.560	0.520
240	0.580	0.530
245	0.590	0.540
250	0.600	0.550
255	0.610	0.560
260	0.620	0.570
265	0.640	0.580
270	0.650	0.590
275	0.660	0.610
280	0.670	0.620
285	0.680	0.630
290	0.700	0.640
295	0.710	0.650
300	0.720	0.660
305	0.730	0.670
310	0.740	0.680
315	0.760	0.690
320	0.770	0.700

Rationale for the use of Ketamine

The goal of anesthesia is to depress the CNS and block the perception of pain. The requirements for reaching a surgical plane of anesthesia are muscle relaxation, absence of reflexes, analgesia, and loss of consciousness. There are numerous methods for achieving this goal. Most of these agents have an additive effect; when combined, less anesthetic is required. It is important to distinguish between these agents. A **sedative** (or tranquilizer) reduces tension or anxiety without affecting physical or mental capabilities. An **analgesic** is an agent which blocks sensation of pain in the conscious state while an **anesthetic** does so in the unconscious state. Many of these agents may have different degrees and combinations of the described effects.

Ketamine cocktail consists of xylazine (a potent sedative and muscle relaxant with analgesic effects), acepromazine (a sedative and anti-emetic), and ketamine (an anesthetic). *Note: Acepromazine has anti-emetic properties. However, rats cannot vomit.* The incorporation of acepromazine allows for a smaller dose of anesthetic and promotes a smoother recovery due to long lasting sedation (several hours). Side effects include suppression of thermoregulation that can lead to hypothermia. Acepromazine is a phenothiazine compound that acts partially by blockage of dopamine receptors in the brain. Xylazine (α-2-adrenergic agonist) is a potent sedative that acts via CNS depression. Xylazine has a significant depressive effect on respiration. In addition, partial atrioventricular block causes a change in cardiac conductivity; thus, decreased heart rate. Its immobilization effects are mediated via inhibition of nerve impulse transmission. Hyperglycemia and diuresis can be side effects. Ketamine is a non-barbiturate dissociative anesthetic that produces effects of immobility and respiratory depression. These are characteristics of the surgical plane (Stage III, Plane 2) of anesthesia that is detailed in a following section.

Ketamine is the author's choice of anesthesia due to its fast, short action, ease of administration, and wide margin of safety. In humans, up to ten times a normal dose has resulted in a longer but complete recovery (Ketamine website). Injectable anesthetic allows for easy access to the head and neck area; eliminating the need for nose cones. Ketamine is highly lipid soluble with a large distribution to fat cells. The metabolism of ketamine is dependent on hepatic clearance. The anesthetic action of ketamine is caused by a disruption between the sensory cortex of the brain and the limbic association areas; hence, the term *dissociative*. The mode of action is thought to include blockage of NMDA (N-methyl-D-aspartate) receptors and the subsequent release of excitatory neurotransmitters. Effects of ketamine include:

- ↑ cardiac output
- ↑ blood pressure and heart rate
- bronchodilation
- salivary secretion
- ↑ cerebral blood flow and pressure
- amnesia
- catalepsy
- open eyes
- normal pharyngeal/ laryngeal reflexes

This cocktail should provide approximately 30 minutes of anesthesia and 60 – 150 minutes of sleep time in the male Sprague Dawley rat. Female rats lack the liver enzyme CYP3A1 and are unable to metabolize as quickly as the males can via CYP3A2. These are the analog equivalents to human CYP3A4; a primary metabolizing enzyme involved in elimination (Kato, 1982). As a result, the anesthesia time for females is the same while the sleep time is considerably longer at 120 – 150 minutes. (Flecknell, 1996, and personal observation). The effects of ketamine and xylazine can be reversed using α2-antagonists such as yohimbine, tolazoline, or atipamazole (Cruz, 1997; Johns Hopkins animal resources website).

An important consideration is the acidic disposition of ketamine. The acid pH is an irritant which can cause soft tissue inflammation, necrosis, and swelling at high doses in rats (Smiler, 1990). This effect can be greatly alleviated when the final concentration is a diluted form and given as multiple injections with a small gauge needle into deep muscle. The author has experienced no complications with acute studies.

Weight Considerations for Surgery

The author recommends a weight range of 230 – 320g (Sprague Dawley rats). **Greater than 320g**: Anesthesia distributes to fat cells where it can temporarily be stored before release into the bloodstream. The heavier rats will store more anesthetic and seem to not be affected. Boosting with small extra doses to extend the surgical plane time, leads to large amounts released into the bloodstream at once and possible overdose due to cumulative effects. Additional complications include difficult dissection through fatty layer, collateral bleeding from the larger sized vessels, and leakage around the catheter. It has been shown that obese rats are more susceptible to infection possibly due to fewer T-cells and lower splenocyte response (Tanaka, 1998). Obese animals exhibit slower wound closure time due to lower wound collagen accumulation in the fat layer (Goodson, 1986; Singer, 1999). **Smaller than 230g**: Rats that are too small or young tend toward poorer survival and longer recovery time. Further complications include anesthetic overdose and small vessels to catheterize.

Reflexes

Absence of reflexes are essential to reaching surgical plane of anesthesia

<u>Pedal Withdrawal Reflex:</u> Extend the hindlimb and pinch the web of skin between the toes causing the foot to withdraw.
<u>Tail Pinch Reflex:</u> Pinch the tail with your fingernails causing the rat to flinch.
<u>Toe Pinch Reflex:</u> Pinch the toe with your fingernails causing the rat to flinch.
<u>Ear Pinna Reflex:</u> Touch the hairs inside the ear canal lightly causing the ear to flick.
<u>Palpebral Reflex:</u> Touch the edge of the eyelid causing the rat to blink.
<u>Corneal Reflex:</u> Lightly touch the edge of the cornea causing the rat to blink

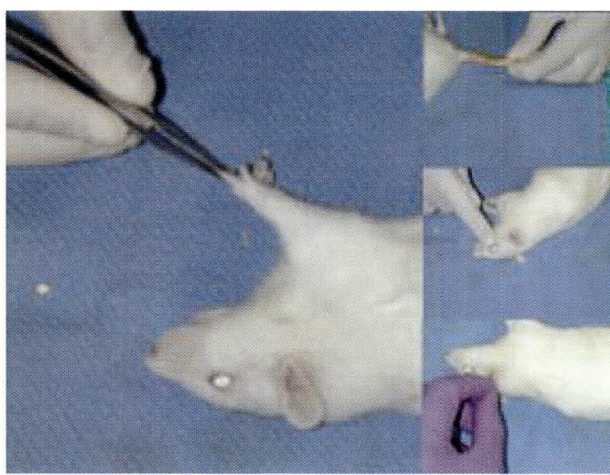

13

Stages of Anesthesia	Definition	Symptom	Mechanism
Stage I	Injection to loss of consciousness	Normal respiration Normal heart rate Ability to vocalize Normal reflexes	
Stage II	Excitatory stage	Delirium Exaggerated sensory responses Excitable activity	Selective depression of higher inhibitory CNS centers
Stage III	Sleep stage	Depressed respiration Depressed circulation Depressed muscle tone Absence of reflexes	
Plane 1	Light anesthesia	Regular pattern of breathing	Mild medullary depression
Plane 2	Medium anesthesia ****Surgical plane****	Irregular breathing Depressed heart rate Depressed blood pressure Strong pulse	Spinal cord depression
Plane 3	Deep anesthesia		Spinal cord depression
Plane 4	Anesthetic overdose		Severe spinal cord depression
Stage IV	Terminal stage	Respiratory distress Cardiac distress Paralysis	
Anesthetic Recovery		Sternal recumbancy Reflexes return in reverse order	

Surgeon's Attire

The surgeon should wear a clean lab coat, disposable jacket, or clean scrub shirt. Wash hands thoroughly with a disinfectant soap. One may wear a clean, fresh pair of lab gloves or sterile gloves. Handling of instruments is easier with a fairly snug fit of gloves. A facemask will help protect the surgeon from reactions to rodent allergens.

Once the surgeon, surgical area, and rat are prepared, one must remain conscious throughout the procedure not to break the aseptic barrier that has been created.

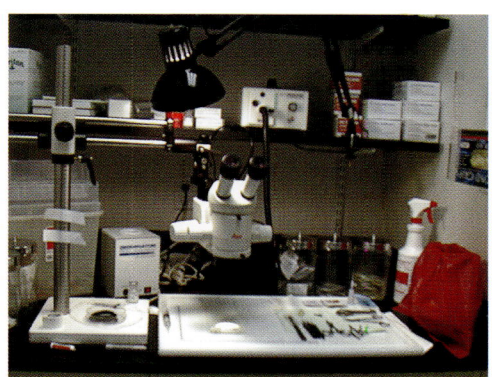

Posture and Muscle Tremor

Correct body posture is extremely important to consider when performing surgery. The surgeon may be required to spend hours in this position each day. Outlined here are some simple recommendations that will help you avoid back and neck pain, muscle fatigue, and muscle tremor. Minute tremors of the hand are greatly amplified through the dissecting scope making accurate, coordinated movements difficult to complete.

1) A well-rested surgeon is best prepared. Weight lifting and coffee drinking before surgery are discouraged since they can exaggerate muscle tremor.

MOTHER GOOSE & GRIMM by Mike Peters

MICRO-SURGERY LAB

STARBUCKS COFFEE

EVENTUALLY THE LAB WAS FORCED TO MOVE

© Grimmy, Inc. Reprinted with special permission of King Features Syndicate

2) Sit in a comfortable lab chair with adjustable back support and adjustable height. Hydraulic tables are commercially available from Boston Tec. The spine along your back and neck should be straight. Both feet should be on the ground slightly apart. A footstool can accomplish this for tall chairs.

3) Adjust the dissecting scope such that your eyes are looking straight ahead and your neck is not bent.

4) Adjust the individual eyepieces to optimize visual clarity and reduce eyestrain. Close one eye and adjust eyepiece to focus clearly. Repeat with other eye. Then use fine focus with both eyes open. Some dissecting scopes allow one to focus without the use of glasses.
5) Rest forearms and elbows on the surgical table.
6) Support the wrists with paper towels or rolled up washcloths to steady the hand. The wrist should always be straight and not flexed. This will allow blood flow and nerve conduction to remain unimpeded. If you experience tingling or numbness in your fingers, drop them below your waist and shake your hand until it goes away. Readjust your hand/wrist position. Remove watches, bracelets, and rings for optimum support and comfort.
7) Do not hold your breath or hyperventilate. Take a deep breath every so often.
8) A radio playing softly in the background can be useful in creating white noise and thereby increasing your concentration by blocking out general and sudden lab sounds.
10) Take a break every 30 – 60 minutes. Stand up, walk around, and stretch.

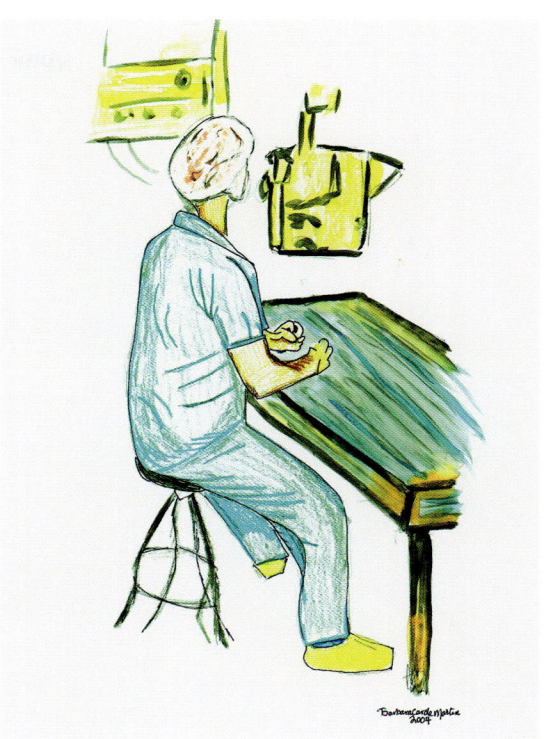

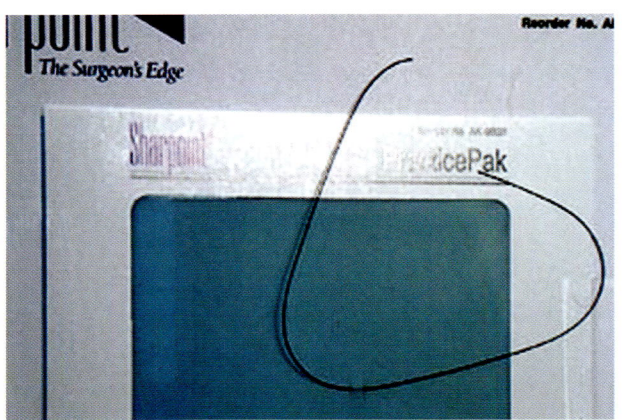

1. Use PracticePak

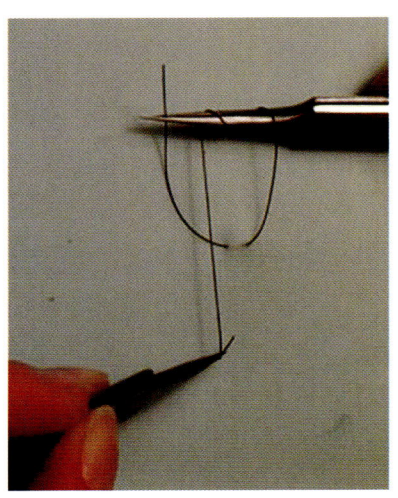

2. Throw 2 loops & grasp other end

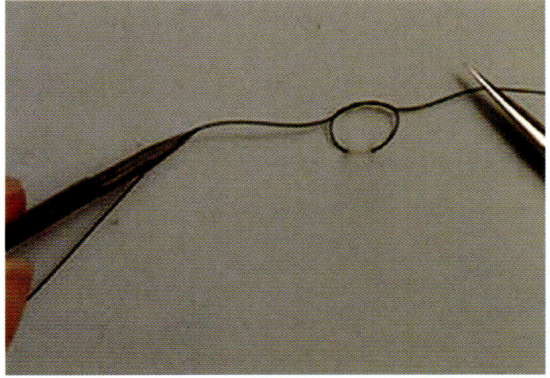

3. Pull through & tighten firmly

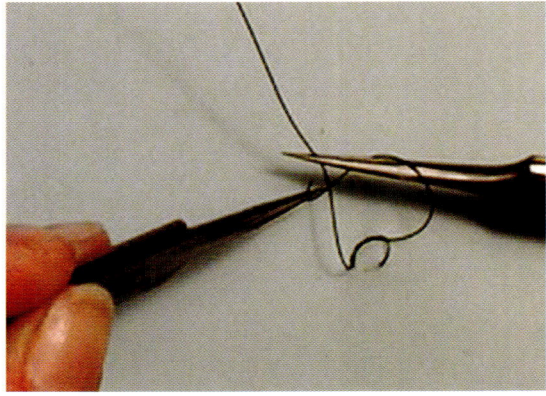

4. Throw 1 loop & grasp other end

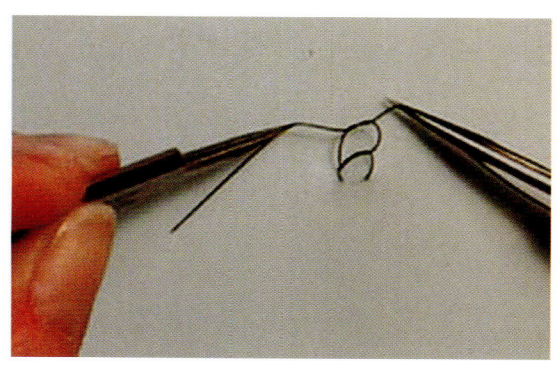

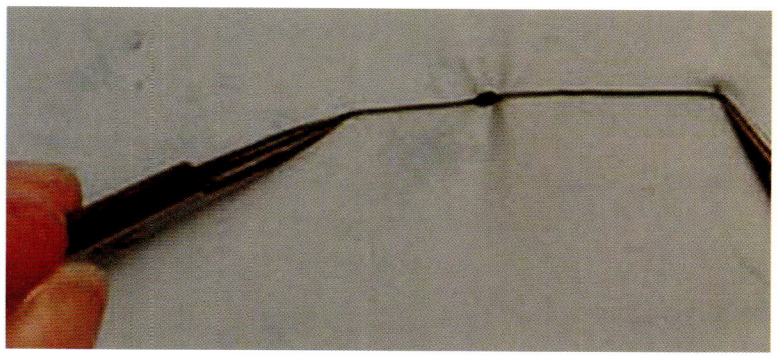

5. Pull through

6. Tighten firmly. The knot is tight enough when no light can be seen through the hole. Tightening too much will constrict blood flow through the catheter.

Hemostasis Techniques

One may employ a number of techniques in order to ensure hemostasis during surgery. Hemostasis is the state of zero blood loss and normal volume in the closed circulatory system. Minimal blood loss during surgery is directly related to increased survival rate. Many factors can contribute to minimal blood loss such as body temperature, blood pressure, ventilation, patient position, and small incision/dissection.

Hypothermic patients tend to lose more blood and use more oxygen. Anesthesia lowers blood pressure, which will slow blood loss. Administration of oxygen during surgery (hyperoxic ventilation) will boost the blood oxygen to vital organs following blood loss.

Several methods for stemming blood loss following incision and dissection include:
- **Pressure** – use sterile gauze or cotton applicators to apply pressure until platelet formation has stopped bleeding.
- **Clamp** – use hemostat, forcep, or vessel clip to clamp at the source of bleeding.
- **Heat** – one may use a sterile or disposable cautery to apply heat in order to coagulate blood.

Recent tools and techniques used in human surgery to control blood loss may be available soon for rodent survival surgery. Coagulating tools include an argon beam plasma coagulator (argon gas and high frequency electrical current), harmonic scalpel (ultrasound waves), and gamma knife (radiation). In addition, laser and cryosurgery techniques minimize blood loss (www.adam.com, Strategies and techniques during surgery). In the case of severe blood loss, replacement fluids may be given intravenously to maintain fluid volume. In humans, following blood donation of one pint (10% blood volume), fluid volume is replaced within 24 hours. Red blood cells, however, may take up to two months to be replenished (www.aabb.org).

CATHETERS

Several types of biocompatible materials are currently used for catheters. Despite this compatibility, infections and blood clots can occur. Infections are primarily caused by catheters contaminated with skin microbes during insertion or the fill solution of the catheter (Goldmann, 1993).

Fill solution

The fill solution usually consists of heparinized saline @ 10 – 1000 IU/mL due to ease of formulation. A heparinized hypertonic glucose solution is an alternative to the use of saline (Mann, 1987). The more viscous additive of heparinized PVPD (polyvinyl pyrrolidone) prevents blood clots and back up of blood into catheter. It is, however, more difficult to handle. Heparinized glycerol is quite viscous though claiming the best patency for long-term applications (Luo, 2000). Heparin-coated catheters have had proven patency of up to 30 days (Foley, 2002).

Stoppers

Stoppers (or obturators) usually consist of wax, nylon filament (fishing line), or stainless steel wire. Bone wax is easier and faster to use while collecting blood. The wax can melt if care is not taken during recovery near a heat source. A solid obturator requires some practice and coordination for quick and efficient use.

Polyethylene catheters

Polyethylene (PE) is the most common catheter material. The use of PE gained popularity following the publication of the Tinsley study in

1983. PE tubing is available in many sizes. Its rigidity helps to prevent twisting/crimping post-implantation and diffusion permeability through the walls. The author prefers this material due to the ease of construction. It is best used for acute studies because it is susceptible to thrombin clots with long term use. It is of note to consider that polyethylene tubing contains toxicants called phthalates. These plasticizers provide characteristics of flexibility and strength but can be absorbed by the animal during longer-term studies. Studies of phthalates in rats report hyperactive thyroid gland following exposure for three months (Price, 1988).

Silicon catheters

Silicon (silastic®) tubing is less likely to form clots. However, the flexibility allows for twisting/crimping post-implantation and can lead to loss of compound as it is absorbed into the walls. The walls will expand easily which can be problematic when attached to continuous infusion pumps, blood pressure apparatus, or automatic blood sampling devices. Correct blood pressure values may not be reflected as a result. Silicon tubing is a good choice of material when using solvents in the vehicle. In general, its inert properties reduce possible vehicle reactions with the tubing material.

Hybrid catheters

Today, it has been shown that hybrid catheters can combine the best attributes of both materials. Silicon is used for implantation while connected to PE tubing for extravasation. These catheters can be laborious to construct and sterilize. They are most advantageous for use in long-term studies and have demonstrated patency for up to two weeks (Arlund, 1997).

Alternative materials

Many companies now offer customized pre-constructed catheters. Braintree Scientific carries a catheter made from a perflourocarbon material which is reported to cause less nerve damage when implanted intrathecally (Sakura, 1996). Other catheter materials include polyvinyl (Tygon®) and polyurethane (Renathane®) which can provide flexibility using natural elastomers. Investigate the possibility of absorption of vehicle and compounds with any delivered fluid.

Catheter Preparation (250g rat)

The 12" carotid artery catheter is marked at 24 mm from the end with a permanent marker. The 12" jugular vein catheter is marked at 12 mm from the implanted end. Make sure the ends of the catheters are cleanly cut without any remaining bevel or jagged edge to avoid unintentional puncture of the vessel walls. Keep a hemostat on the carotid catheter until ready to flush back. Do not allow jugular catheter to fall below table height. Gravity will cause blood to escape even with low venous pressure.

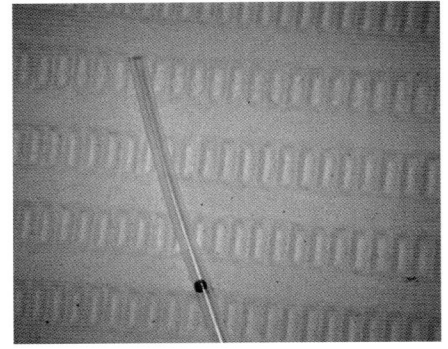

General Dissection

Figure 1 displays the underlying musculature. Figure 2 demonstrates the complex in situ anatomy of the vessels in the neck of the rat. These figures should assist in the gross dissection and location of the carotid artery and jugular vein.

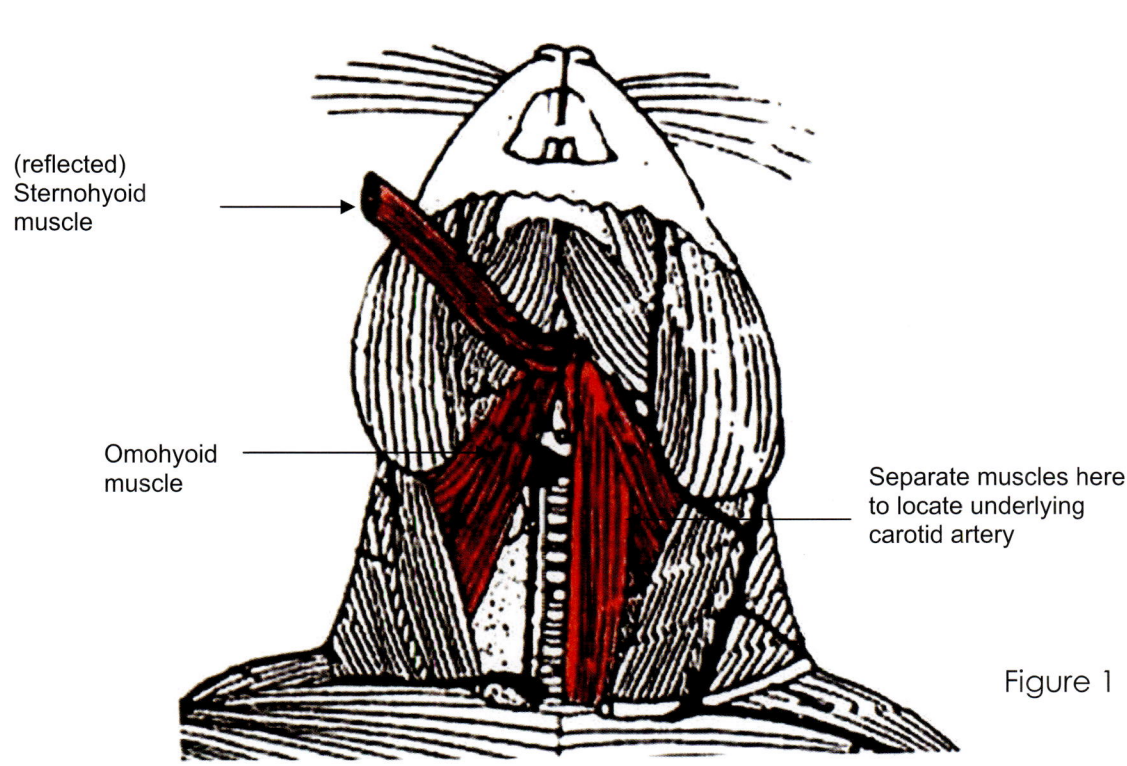

(reflected)
Sternohyoid
muscle

Omohyoid
muscle

Separate muscles here
to locate underlying
carotid artery

Figure 1

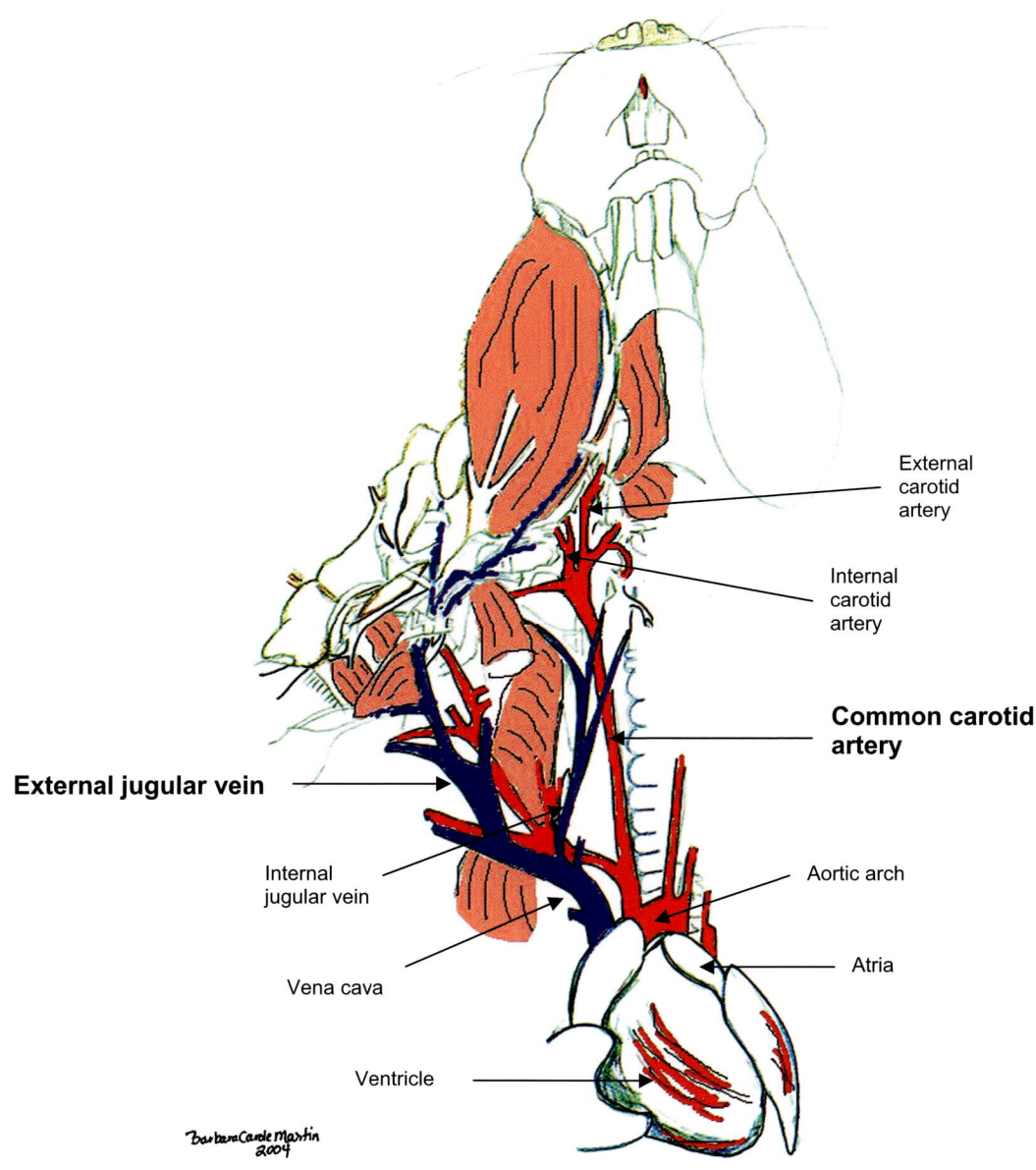

External
carotid
artery

Internal
carotid
artery

**Common carotid
artery**

External jugular vein

Internal
jugular vein

Aortic arch

Atria

Vena cava

Ventricle

Barbara Carole Martin
2004

Figure 2

30

Vessel Anatomy (CAROLINA BIOLOGICAL SUPPLY COMPANY, PICTURE USED WITH PERMISSION)

The differences between arteries and veins can be described as such: Arteries are more flexible and carry blood away from the heart at relatively high pressure. Veins are less flexible due to fewer elastic fibers and contain semi-lunar valves which allow blood to flow only towards the heart at a lower pressure. Blood vessels, with the exception of capillaries, contain three distinct layers in the vessel wall. The innermost layer in all vessels is the *tunica intima* or endothelium. Endothelium cells can detect pressure, oxygen, and flow changes. They have the ability to affect vascular smooth muscle tone. Oxygen is obtained by diffusion from red blood cells in this layer. The middle layer consists of smooth muscle and autonomic nerves. It is referred to as the *tunica media*. This layer is very elastic and mostly in arteries. The outermost layer, called *tunica adventia*, is the most prominent in veins. It is comprised of primarily collagen, some smooth muscle, and autonomic nerves.

One must be mindful of these properties when manipulating the carotid artery and jugular vein. Excessive handling can cause vaso-constriction and shredding of the vessel walls. The more delicate, fibrous layers of the jugular vein can separate making insertion of the catheter difficult and frustrating. Care must be taken in the placement and location of the catheter. Waynforth and Flecknell (1992) describe the descending aortic arch and the entrance to the right atrium as the optimum placement for the carotid and jugular catheters, respectively. The author prefers a more distal placement to the heart to accommodate the variability in body size. The distance of insertion should be adjusted to body size (1 cm/100g BW) if the Waynforth placement is desired.

31

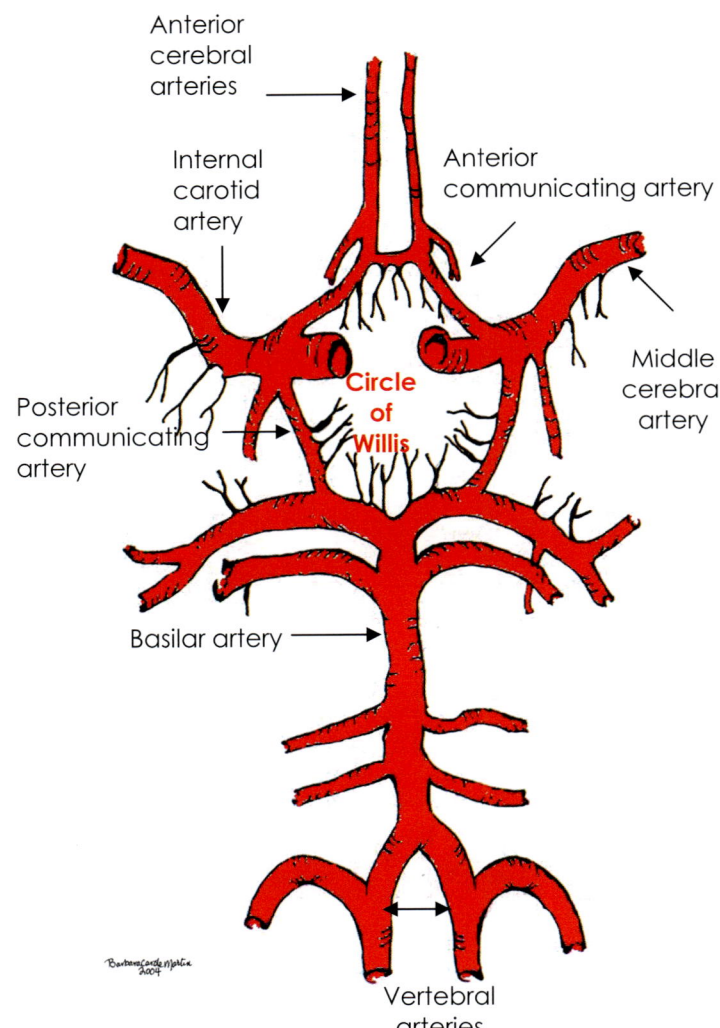

Anterior
cerebral
arteries

Internal
carotid
artery

Anterior
communicating artery

Posterior
communicating
artery

Circle
of
Willis

Middle
cerebral
artery

Basilar artery

Barbara Cousle Merle
2004

Vertebral
arteries

Unilateral Carotid Occlusion: Blood Flow

The carotid and vertebral arteries on both sides of the body provide the blood supply to the brain. The circulation from both sides is connected by the circle of willis in the brain. The circle of willis allows both sides of the brain and body to maintain pressure equilibrium. Collateral circulation to other organs travels from the carotid via the ipsilateral pterygopalatine, superior thyroid, and occipital arteries (Smith, 1996). In humans, these connecting vessels have too small a diameter to maintain adequate blood supply following unilateral carotid occlusion. In many other mammals, the anterior and posterior communicating vessels are abnormal or missing. Gerbils have unique anatomy; up to one third have no anterior communicating vessels. These animals experience severe stroke when a unilateral carotid occlusion is performed (Davson, 1996). The rat communicating vessels are present and can carry sufficient supply of blood to prevent stroke when unilateral carotid occlusion is performed. In addition, rats have been shown to develop extra collaterals when the carotid is chronically occluded (Coyle, 1990).

Vagus Presentation

In a normal presentation, the vagus nerve runs parallel to the carotid artery for the length exposed.

In a crossover presentation, the vagus nerve crosses the carotid artery at unpredictable locations.

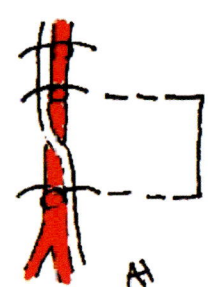

Choose your suture placement carefully to avoid possible trauma and damage to the vagus nerve caused by friction against the suture material. Vagal stimulation can result in cardiac abnormalities or failure.

ANTERIOR

33

Practice Surgery

"Retrospective reviews ... clearly indicate that operative experience is the single most critical factor related to improved success rates. Surgeons with a high success rate went through a learning curve, and most of them still learn from occasional complications. They improve by learning what they ... realized through hindsight to be a mistake." ~ Khouri, 1992

"The difference between one experimenter's success with a technique and another's failure may have as much to do with differences in the experimenter's skill as with any flaws inherent in the technique." ~ Giner, 1987

PracticePak Practice
Practice suture techniques, knot tying, or getting used to microscopic surgery and posture using a Sharpoint suture pak. This pak is designed by Surgical Specialties to resemble the texture of skin and is invaluable for orienting the surgeon before using live animals or cadavers.

Rat Model Practice
Microsurgical Developments PVC-Rat from Braintree Scientific is useful for practice of both catheterizations.

Cadaver Practice
It is advisable to use a rat cadaver prior to using a live animal. Cadavers will assist in the orientation of the anatomy and insertion of the catheter without the complication of blood pressure.

Non-survival Practice

The final step for the beginner is to try a live animal, under a more experienced surgeon's supervision, and sacrifice the rat before recovery from anesthetic. This is an excellent way to practice and reduce the time in which a catheterization can be performed. Ketamine cocktail can be boosted 100μl every 30 minutes. Several drops of lidocaine on the artery can reduce vasospasms and vasoconstriction, which are evident following the excessive manipulation during practice surgeries. Less procedure time will decrease the stress to the rat and reduce recovery time. The first few jugular and carotid catheterizations will generally take 30 – 60 minutes per rat. The average time in which a proficient surgeon should be able to complete the two procedures is 12 – 15 minutes per rat.

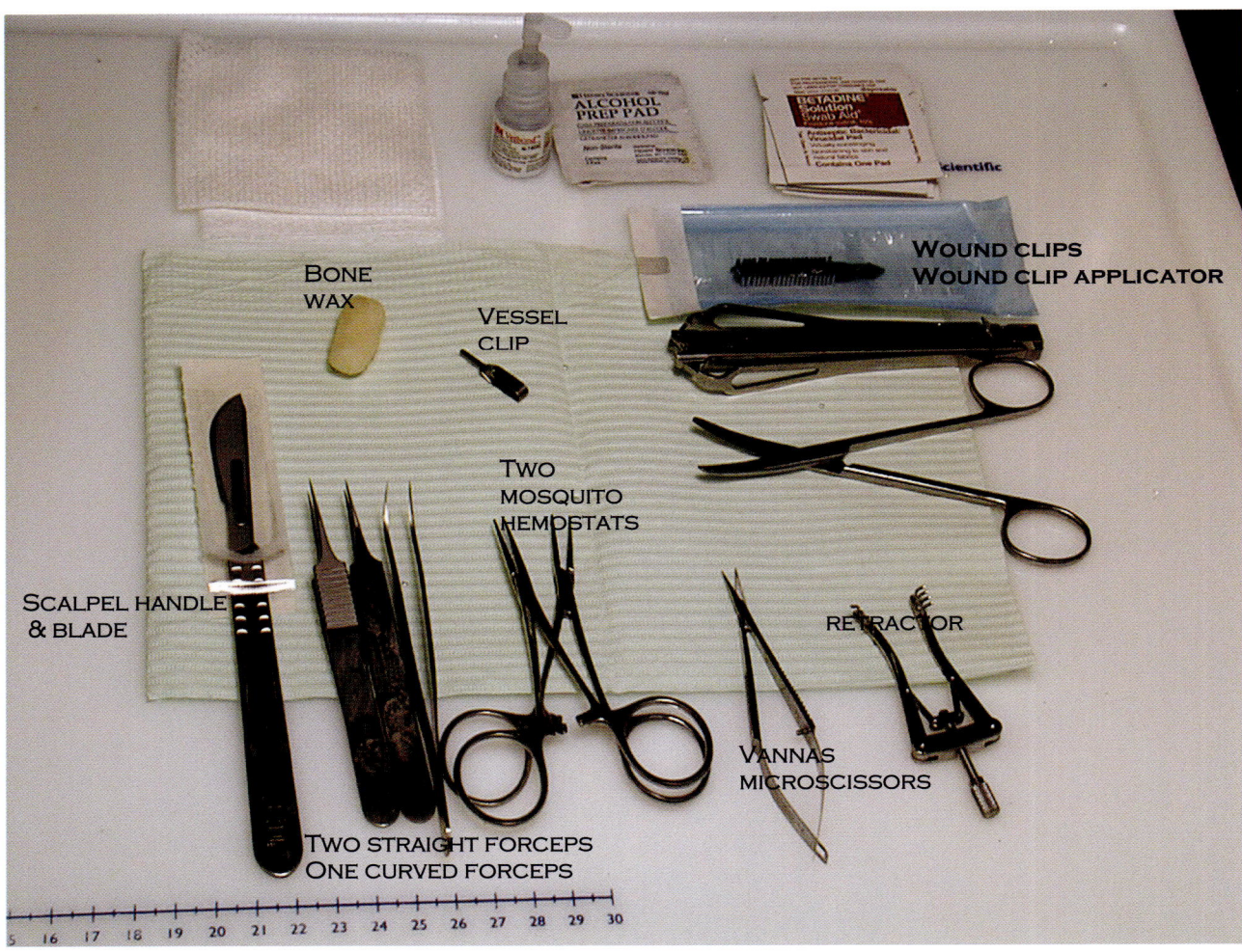

Bone wax

Vessel clip

Wound clips
Wound clip applicator

Two mosquito hemostats

Scalpel handle & blade

Retractor

Vannas microscissors

Two straight forceps
One curved forceps

36

JUGULAR
VEIN
CATHETERIZATION

NOTE: IN THE FOLLOWING
PICTURES, THE INCISION
SIZE IS LARGER AND
DRAPES ARE ABSENT FOR
A CLEARER PRESENTATION

MATERIALS	PROCEDURE	PICTORIAL

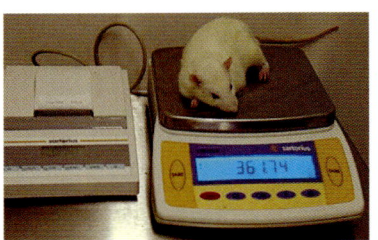

MATERIALS

SCALE
GLOVES
PERMANENT MARKER

DECAPICONE

(INSULIN
SYRINGE+28G½
NEEDLE)
KETAMINE COCKTAIL:
KETAMINE
XYLAZINE
ACEPROMAZINE
STERILE SALINE

HEATING PAD

PROCEDURE

PUT ON GLOVES.
WEIGH RAT & RECORD.
MARK RAT ID# ON BASE
OF TAIL.

RESTRAINT

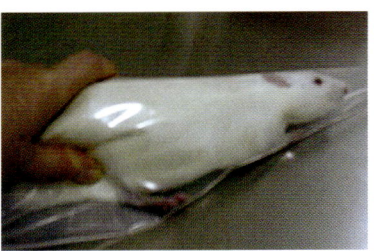

IM INJECTION @
2.2 – 2.4 ML/KG
HALF EACH THIGH
USING RESTRAINT.

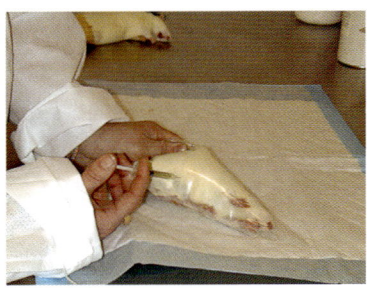

MAINTAIN BODY HEAT
BEFORE, DURING, & AFTER
SURGERY
ON LOW SETTING.

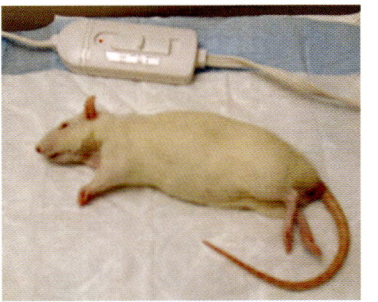

MATERIALS	PROCEDURE	PICTORIAL

MATERIALS

HOT BEAD STERILIZER
INSTRUMENTS
STERILE TOWEL

PROCEDURE

STERILIZE INSTRUMENTS.

PICTORIAL

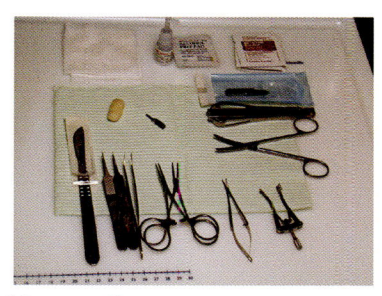

NOTE: STERILIZE TOWELS, GAUZE, WOUND CLIPS, SUTURES & COTTON APPLICATORS

CLIPPERS

SHAVE HAIR ON ANTERIOR UPPER THORAX (A) & OVER SCAPULA (B).

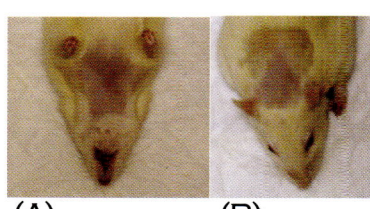

(A)   (B)

SURGICAL BOARD
TAPE
GAUZE BOLSTER:
  ROLL UP TWO PIECES
  OF 3"X3" GAUZE AND
  TAPE TOGETHER

IN DORSAL RECUMBENT POSITION, SECURE FORELIMBS WITH TAPE. PLACE BOLSTER UNDER NECK. TO CLEAR AIRWAY, PULL OUT TONGUE TO SIDE. CHECK REFLEXES.

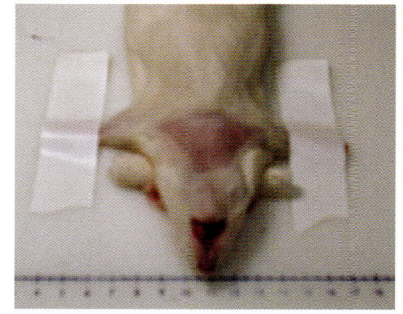

MATERIALS

BETADINE PAD
ALCOHOL PAD

STERILE DRAPE OR
3"x3" STERILE GAUZE

NO. 4 SCALPEL HANDLE
& NO.22 BLADE

STERILE 3" x 3" GAUZE
STERILE COTTON
  APPLICATOR

PROCEDURE

WIPE BETADINE IN
CONCENTRIC CIRCLE
FROM MIDDLE TO OUTSIDE.
WIPE WITH ALCOHOL PAD.
REPEAT TWICE MORE.
WIPE SCALPEL BLADE
WITH USED PAD (HEAT
DULLS BLADE).

DRAPE INCISION SITE.

MAKE 2 – 3 CM MIDLINE
SKIN INCISION WHILE
STRETCHING SKIN TO
CREATE TENSION.

STOP ANY BLEEDING.

PICTORIAL

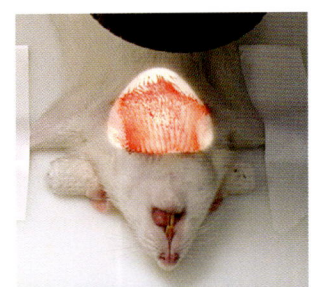

 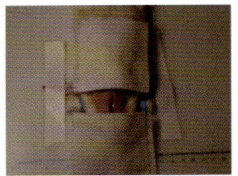

NOTE:
IF CATHETERIZING ONLY THE
JUGULAR VEIN, MAKE A SMALLER
INCISION CLOSER TO THE VEIN

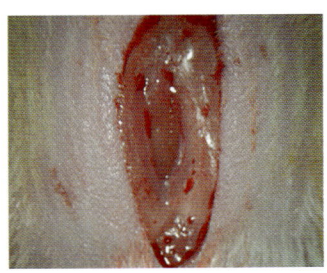

40

MATERIALS	PROCEDURE	PICTORIAL

MATERIALS

STRAIGHT FORCEP
MOSQUITO HEMOSTAT

PROCEDURE

LOCATE RAT'S LEFT JUGULAR VEIN BY LIFTING UP SKIN WITH FORCEP & SEPARATE FROM FAT LAYER WITH HEMOSTAT USING A SCISSORS ACTION.

OBSERVE THE EXPOSED JUGULAR VEIN.

STRAIGHT FORCEP
CURVED FORCEP

GRAB CONNECTIVE TISSUE AT ANTERIOR END WITH STRAIGHT FORCEPS (1). USING THE CURVED FORCEPS (2) IN A SCISSORS ACTION, MAKE SMALL POCKET ON EACH SIDE OF THE JUGULAR.

PICTORIAL

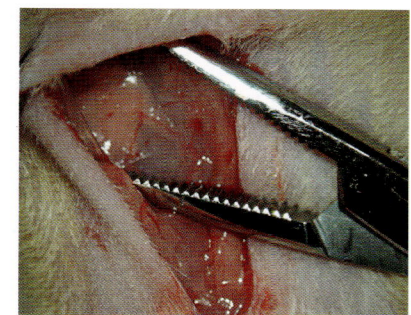

NOTE: HEMOSTAT CAN BE USED TO RETRACT SKIN. REDUCE TRAUMA IF POSSIBLE.

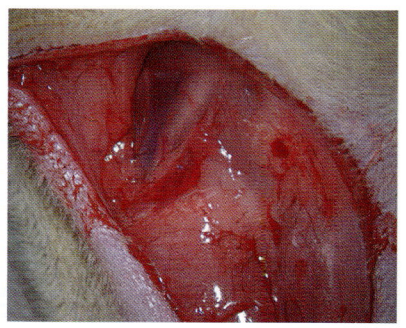

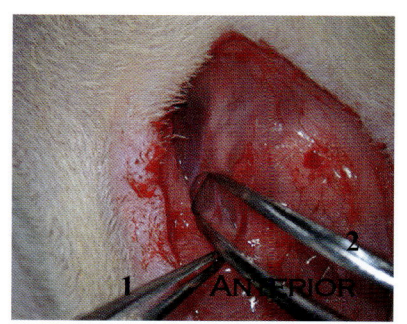

MATERIALS	PROCEDURE	PICTORIAL
STRAIGHT FORCEP CURVED FORCEP	PLACE CLOSED CURVED FORCEP (2) PARALLEL TO JUGULAR VEIN, POINT DOWNWARDS, & THEN POINT IT PERPENDICULAR TO THE VEIN & PUSH UNDERNEATH.	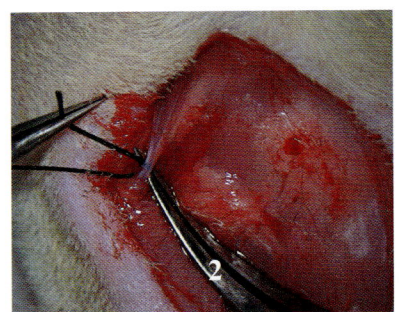
STERILE 6-INCH SILK SUTURE	OPEN FORCEPS, GRAB THE SUTURE & PULL BACK THROUGH. THROW TWO LOOPS & PULL TIGHTLY.	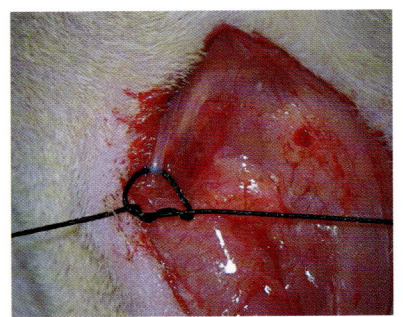
	THROW ONE LOOP IN OTHER DIRECTION & PULL TIGHTLY.	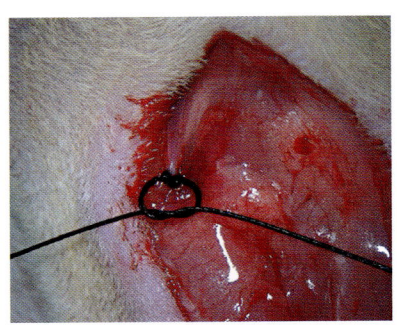

MATERIALS	PROCEDURE	PICTORIAL

MOSQUITO HEMOSTAT

ATTACH HEMOSTAT TO LOOSE ENDS FOR TENSION.

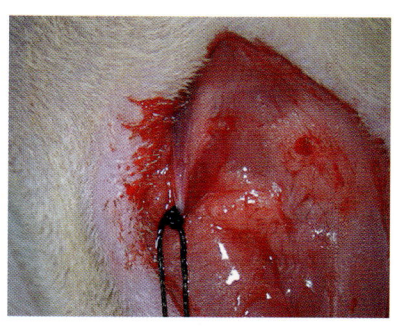

STRAIGHT FORCEP
CURVED FORCEP

USE FORCEPS TO MAKE POCKETS AT POSTERIOR END OF EXPOSED VEIN.

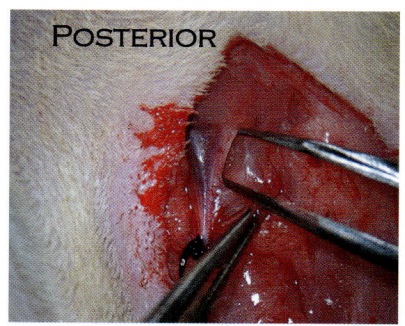

CAREFULLY PUSH CURVED FORCEP UNDERNEATH AVOIDING PUNCTURE OF VEIN.

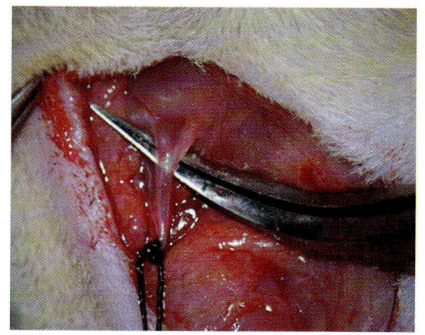

43

MATERIALS	PROCEDURE	PICTORIAL

MATERIALS

STERILE 6-INCH SUTURE
MOSQUITO HEMOSTAT

PROCEDURE

PULL ANOTHER SUTURE LENGTH THROUGH. THROW TWO LOOPS LOOSELY & ATTACH HEMOSTAT TO APPLY TENSION (A). DO NOT TIE OFF.

PICTORIAL

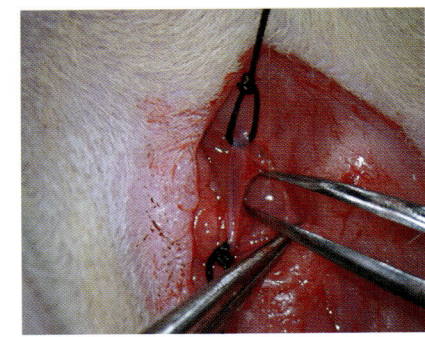

TIP: KEEP LOOSE SUTURE END TAUT WHILE PULLING THROUGH TO REDUCE FRICTION TO VEIN ·

STRAIGHT FORCEP
CURVED FORCEP

USE FORCEPS TO MAKE POCKETS AT MIDDLE OF EXPOSED VEIN.

STRAIGHT FORCEP
CURVED FORCEP

PUSH CURVED FORCEP UNDERNEATH. USE STRAIGHT FORCEP TO REFLECT SKIN.

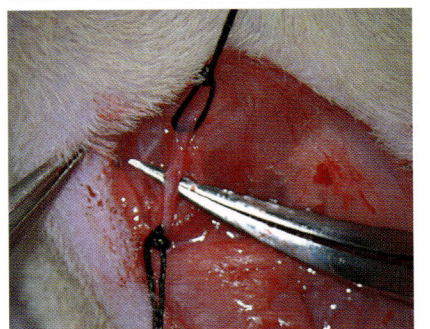

MATERIALS	PROCEDURE	PICTORIAL

STERILE 6-INCH SUTURE

GRAB SUTURE LENGTH & PULL THROUGH WITHOUT ANY LOOPS OR TIES.

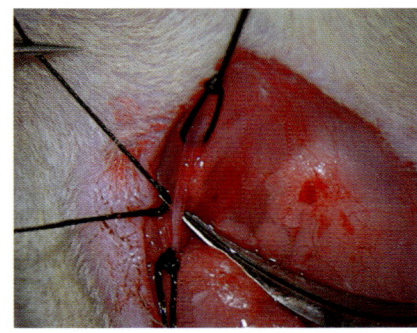

VANNAS MICROSCISSORS

MAKE SMALL "V" CUT (VENOTOMY) INTO TOP OF VEIN. TENSION ON THE POSTERIOR SUTURE SHOULD PREVENT BLOOD FLOW. INCREASE TENSION IF NECESSARY.

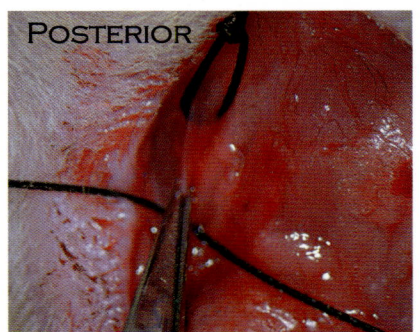

OBSERVE OPENING.

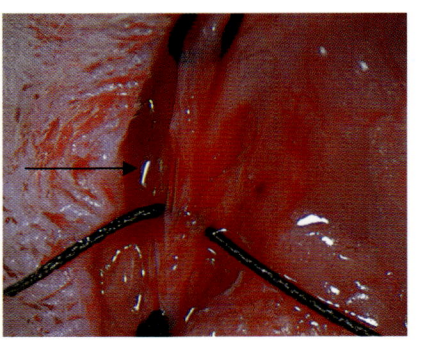

MATERIALS

TWO STRAIGHT FORCEPS
12 INCH PE50 (MARKED
@ 12MM WITH MARKER)
ATTACHED TO 22½G
NEEDLE ON 10 ML
SYRINGE FILLED WITH
100IU HEPARINIZED
SALINE

PROCEDURE

LIFT UP FLAP OF "V" TO
OPEN HOLE FOR EASIER
INSERTION.

INSERT BLUNT TIPPED
CATHETER INTO HOLE
UNTIL MARK IS AT THE
HOLE.

OBSERVE LOCATION.

PICTORIAL

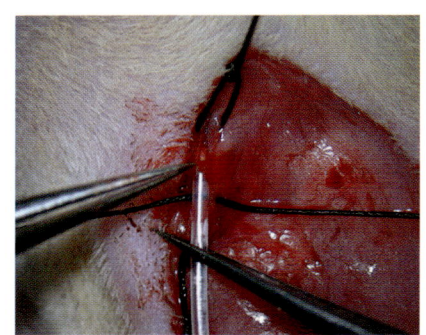

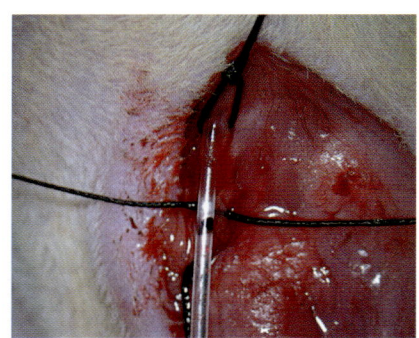

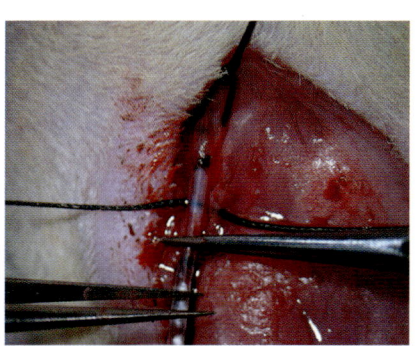

MATERIALS

TWO STRAIGHT FORCEPS

PROCEDURE

TIE OFF SUTURE (1) WITH SURGEON'S KNOT TO SECURE VENOTOMY SITE.

KEEP SUTURES STRAIGHT TO AVOID CONFUSION WHILE TYING OFF. TIE OFF SUTURE (2) WITH ADDITIONAL THROW IN OPPOSITE DIRECTION.

TIE OFF SUTURE (3) TO SECURE VEIN TO CATHETER AS ANCHOR. THREE SUTURES PROVIDE SUPERIOR SECURITY.

PICTORIAL

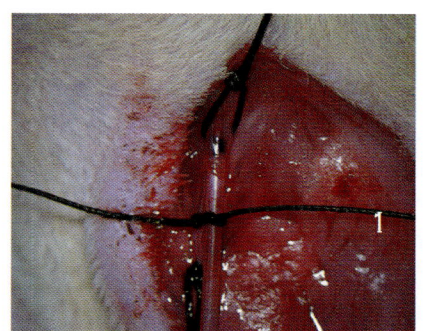

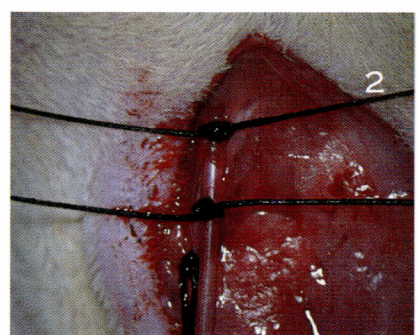

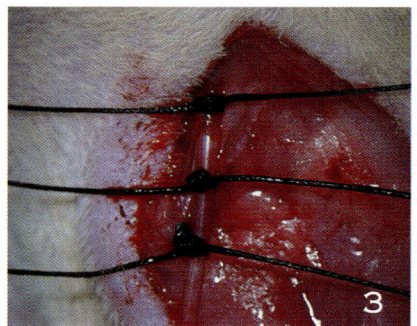

MATERIALS

LARGE SCISSORS

10CC SYRINGE FILLED
  WITH 100IU HEPARIN/
  ML SALINE

STERILE BONE WAX

PROCEDURE

CUT EXTRA LENGTH OFF.
DO NOT CUT TOO SHORT
TO AVOID UNRAVELING.
DO NOT LEAVE TOO LONG
TO AVOID WICKING.

CHECK SUCCESS BY
WITHDRAWING BLOOD &
THEN FLUSH BACK. MAKE
SURE CATHETER IS CLEAR
OF BLOOD TO AVOID
CLOTS.

REMOVE TUBING FROM
NEEDLE & PLUG WITH
STERILE BONE WAX.

PICTORIAL

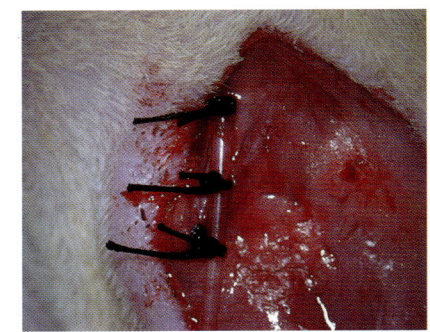

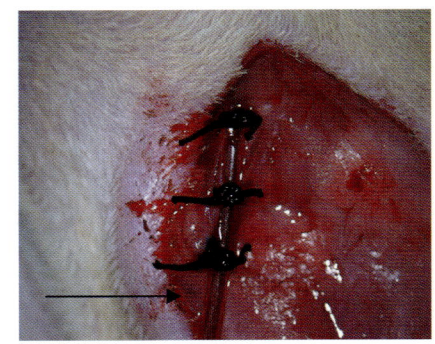

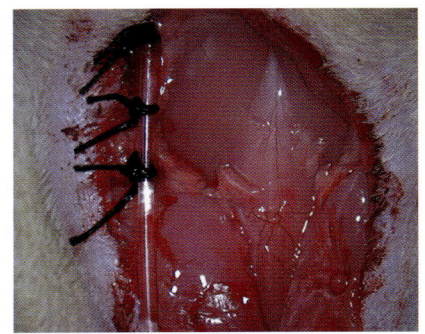

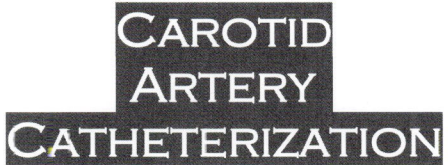

CAROTID ARTERY CATHETERIZATION

NOTE:
ONE MAY CATHETERIZE THE CAROTID ARTERY FOLLOWING THE JUGULAR VEIN CATHETERIZATION <u>OR</u> CATHETERIZE THE CAROTID ARTERY ONLY. FOR CAROTID ONLY, FOLLOW THE PREPARATORY STEPS OUTLINED FOR THE JUGULAR CATHETERIZATION AND MAKE A SMALLER MIDLINE INCISION.

MATERIALS

Mosquito hemostat
Straight forcep

PROCEDURE

Locate omohyoid muscles. Reflect salivary glands to side. With scissors action of hemostat, separate right omohyoid from the sternohyoid muscle.

Use forceps to provide tension, if necessary, to expose the common carotid artery.

PICTORIAL

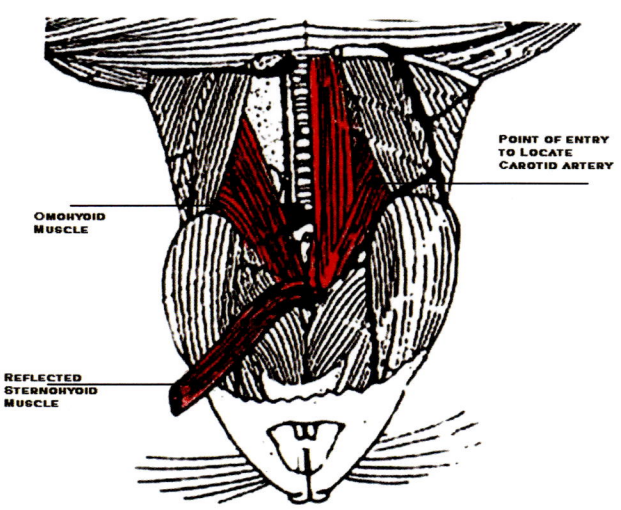

(Jugular vein catheter is to the left)

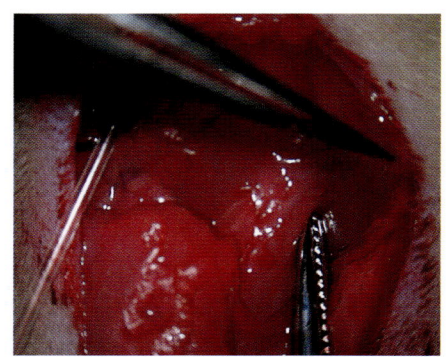

MATERIALS

RETRACTOR
  IN CLOSED POSITION

#5 STRAIGHT FORCEP
VANNAS
  MICROSCISSORS

PROCEDURE

INSERT AND OPEN
RETRACTOR TO SECURE
CLEAR VIEW OF RIGHT
COMMON CAROTID
ARTERY. CHECK
RESPIRATION DUE TO
PRESSURE ON TRACHEA.

CUT AWAY ANY RAGGED
EDGES TO CLEAR VIEW.

LIFT REMAINING MUSCLE
AWAY FROM ARTERY AND
CAREFULLY TRANSECT.

PICTORIAL

NOTE: AVOID PUNCTURE OF
ESOPHAGUS OR TRACHEA WITH
RETRACTOR

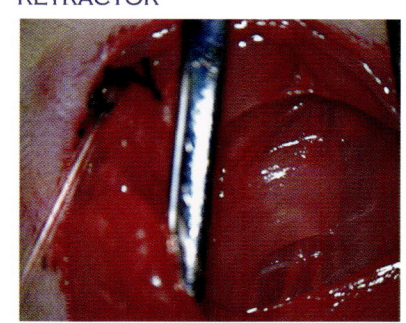

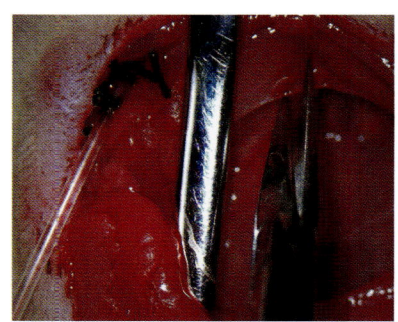

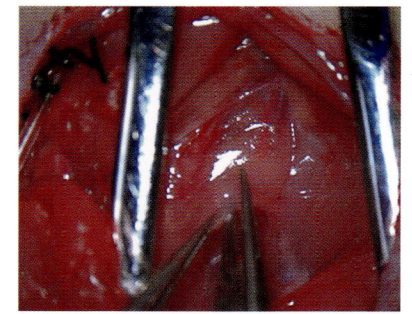

51

MATERIALS

CURVED FORCEPS
STRAIGHT FORCEPS

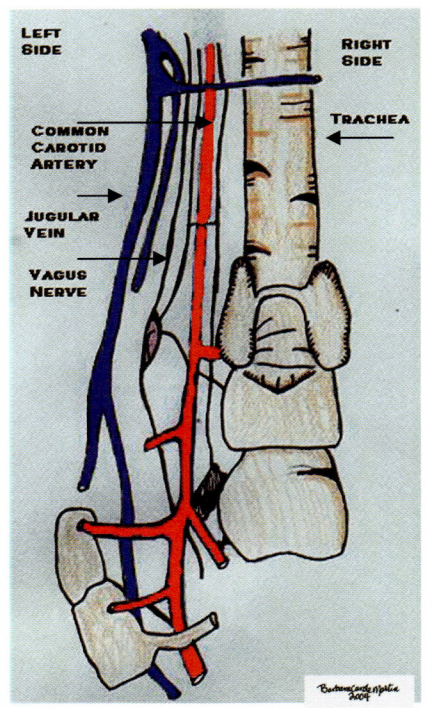

PROCEDURE

GRAB CONNECTIVE TISSUE AT ANTERIOR END WITH STRAIGHT FORCEPS & PULL AWAY SLIGHTLY FROM CAROTID TO LOCATE THE VAGUS NERVE.

USING THE CURVED FORCEPS IN A SCISSORS ACTION, MAKE SMALL POCKET BETWEEN THE ARTERY & NERVE, POSTERIOR TO THE BIFURCATION OF EXTERIOR & INTERIOR CAROTID, WITHOUT TOUCHING THE VAGUS.

PICTORIAL

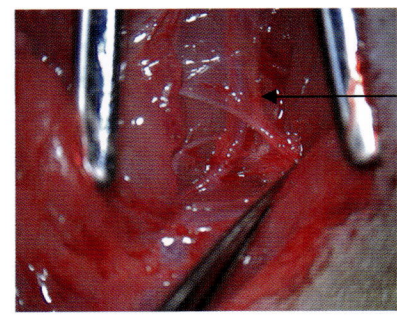

ANTERIOR

NOTE: DO NOT TOUCH VAGUS NERVE. IT CONTROLS BREATHING. THE RECOVERED RAT WILL GASP. THE SYMPATHETIC NERVE CONTROLS PUPIL SIZE. IF CUT, THE RECOVERED RAT WILL SQUINT IN THAT EYE.

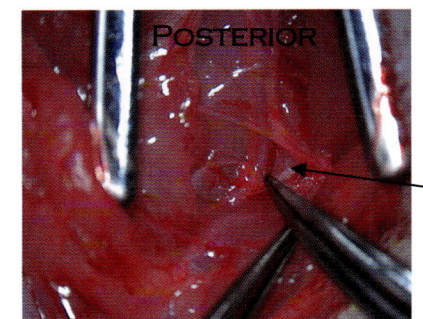

MATERIALS	PROCEDURE	PICTORIAL

CURVED FORCEPS
STRAIGHT FORCEPS

PLACE CLOSED CURVED FORCEP PARALLEL TO THE MIDDLE OF THE CAROTID, POINT DOWNWARDS INTO POCKET, & THEN POINT IT PERPENDICULAR TO THE ARTERY & PUSH UNDERNEATH.

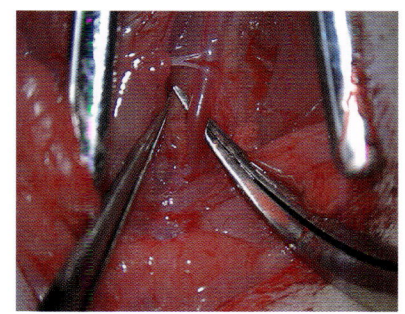

STERILE 6-INCH SUTURE

OPEN FORCEPS, GRAB A SUTURE LENGTH & PULL BACK THROUGH.

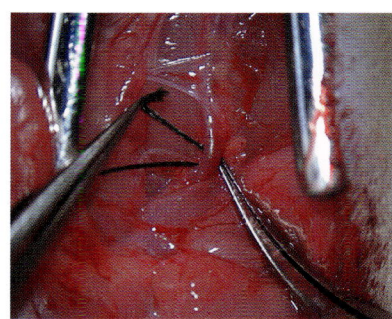

HEMOSTAT

TIE A SURGEON'S KNOT & SECURE WITH HEMOSTAT TO APPLY TENSION.

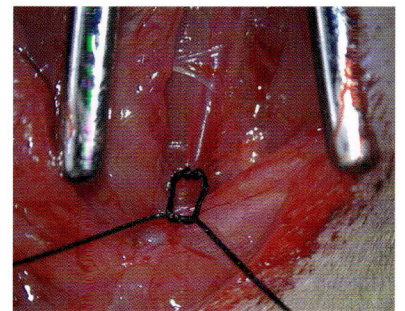

MATERIALS

CURVED FORCEPS
STRAIGHT FORCEPS

PROCEDURE

USE FORCEPS TO MAKE
POCKETS AT MIDDLE
BETWEEN VAGUS &
ARTERY.

PUSH CURVED FORCEP
UNDERNEATH VESSEL
TAKING CARE TO NOT
INCLUDE VAGUS.

STERILE 6 INCH SUTURE

PULL ANOTHER SUTURE
LENGTH THROUGH.

PICTORIAL

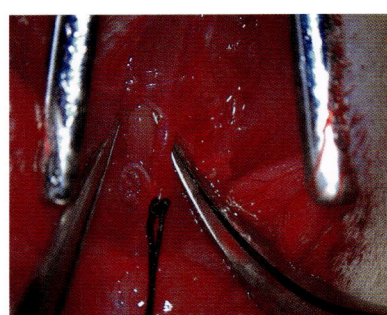

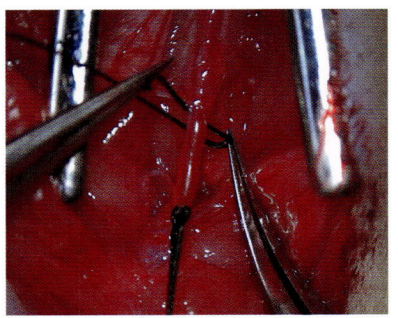

54

MATERIALS

CURVED FORCEPS
STRAIGHT FORCEPS

PROCEDURE

THROW TWO LOOPS.
LEAVE LOOSE.

PICTORIAL

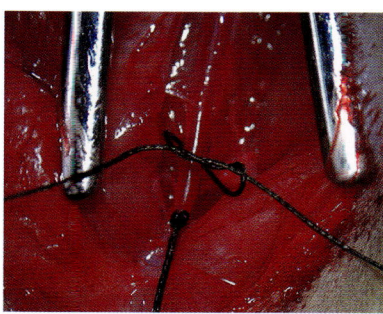

TIP: KEEP SUTURE LENGTHS
STRAIGHT & UNTANGLED

STERILE 6-INCH SUTURE

REPEAT POCKET AND
LOOSE SUTURE AT
POSTERIOR END.

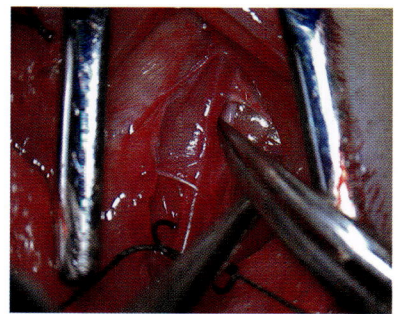

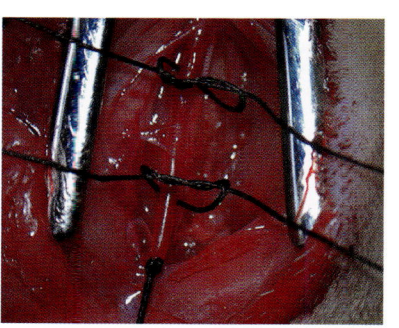

MATERIALS

VESSEL CLIP
VANNAS
    MICROSCISSORS
TWO STRAIGHT FORCEPS

12 INCH PE50 MARKED
@ 24MM WITH MARKER
ATTACHED TO 22 1|G
NEEDLE ON 10 mL
SYRINGE FILLED WITH
100IU HEPARINIZED
SALINE — <u>CLAMP
CATHETER WITH
HEMOSTAT AT
UNMARKED END TO
PREVENT BLOOD FLOW</u>

PROCEDURE

PLACE VESSEL CLIP INTO
POSTERIOR POCKETS TO
STOP BLOOD FLOW
FROM HEART. DO NOT
CLIP VAGUS. USE
STRAIGHT FORCEP TO
OPEN POCKET FOR CLIP.

MAKE V-CUT INTO TOP
OF ARTERY BETWEEN 2ND
AND 3RD POSTERIOR
SUTURES. SOME
TRAPPED BLOOD WILL
"POP" OUT. COAT
DRYING VESSEL WITH
THIS BLOOD.

THREAD BLUNT
CATHETER INTO HOLE UP
TO CLIP.

PICTORIAL

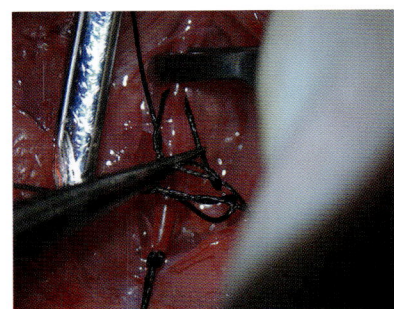

TIP: LIFT UP FLAP OF "V" WITH
STRAIGHT FORCEPS CREATING A "TENT"
TO THREAD CATHETER

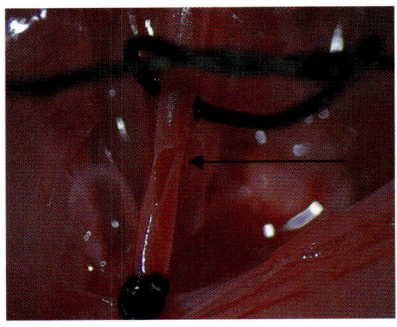

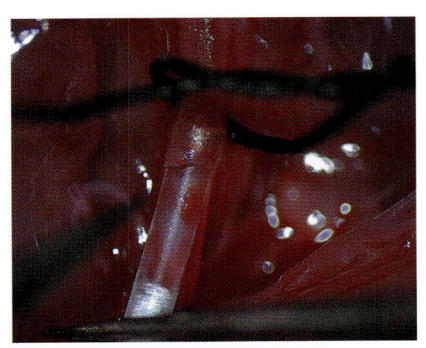

MATERIALS

TWO STRAIGHT FORCEPS

VESSEL CLIP

PROCEDURE

HOLD FORCEP 1 AROUND ARTERY & CATHETER TIGHTLY JUST POSTERIOR TO V-CUT SO NO BLOOD WILL ESCAPE. REMOVE CLIP.

USING FORCEP 2, PUSH CATHETER INTO ARTERY UNTIL MARK IS AT POSTERIOR SUTURE.

WHILE STILL HOLDING ARTERY CLOSED WITH FORCEP 1 DROP FORCEP 2 & QUICKLY PLACE VESSEL CLIP BACK IN SAME PLACE OVER CATHETER.

PICTORIAL

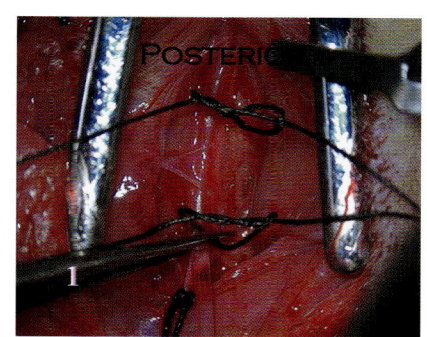

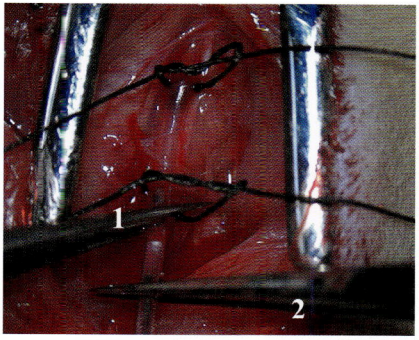

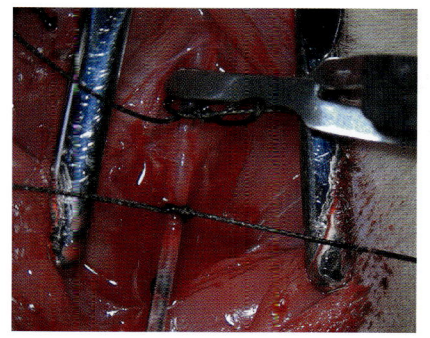

MATERIALS	PROCEDURE	PICTORIAL

TWO STRAIGHT FORCEPS

TIE TWO POSTERIOR SUTURES & REMOVE VESSEL CLIP. TIE ANTERIOR SUTURE.

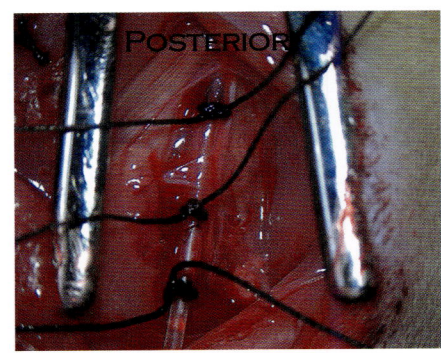

SCISSORS

CUT EXCESS LENGTH.

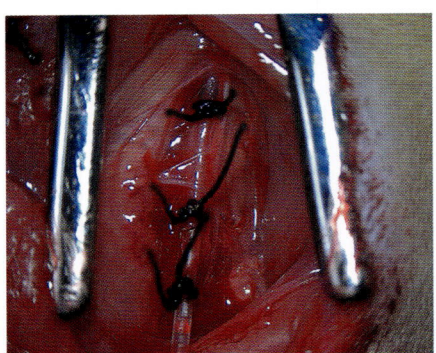

REMOVE RETRACTOR.

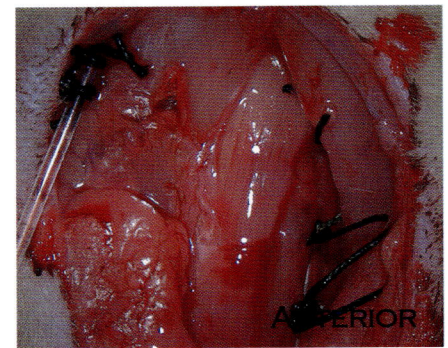

58

MATERIALS	PROCEDURE	PICTORIAL

STERILE BONE WAX

UNCLAMP HEMOSTAT FROM CATHETER END & WITHDRAW BLOOD TO CHECK SUCCESS. FLUSH BACK UNTIL CLEAR. RE-CLAMP & DISCONNECT NEEDLE. PLUG END WITH STERILE BONE WAX. UNCLAMP HEMOSTAT.

TIP: DO NOT OVER FLUSH. EXCESSIVE HEPARIN WILL CAUSE INTERNAL BLEEDING.

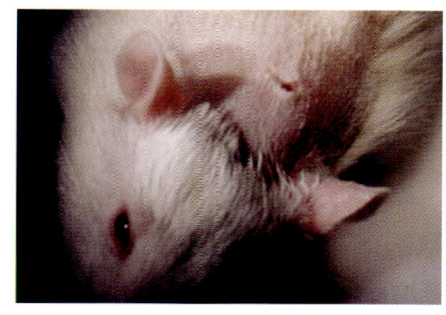

SCISSORS

UNTAPE & TURN RAT ONTO SIDE. MAKE SMALL CUT ON BACK OF NECK.

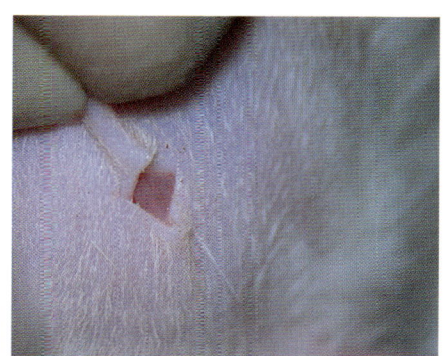

MATERIALS

HEMOSTAT

PROCEDURE

STAYING CLOSE TO THE
SKIN, INSERT CLOSED
HEMOSTAT @ BACK OF
NECK & PUSH THROUGH
TO VENTRAL INCISION.
OPEN HEMOSTAT &
GRAB CATHETER.

PULL JUGULAR
CATHETER THROUGH
LEFT SIDE TO BACK OF
NECK. PULL CAROTID
CATHETER THROUGH
RIGHT SIDE TO BACK OF
NECK. THE TRACHEA
SHOULD NOT BE
CROSSED.

PICTORIAL

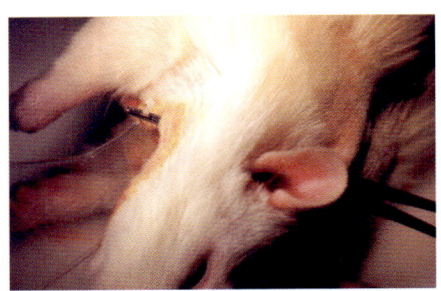

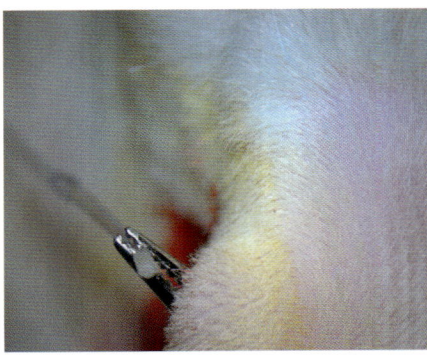

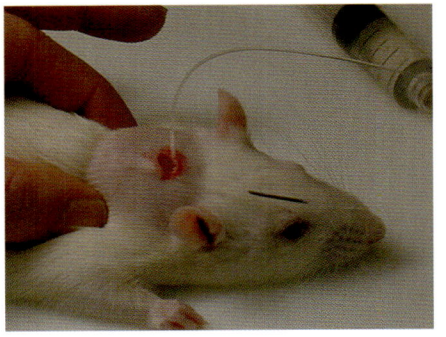

MATERIALS

WOUND CLIPS
WOUND CLIP APPLIER

PROCEDURE

PULL SKIN EDGES
TOGETHER UNDER NECK.

APPLY WOUND CLIPS TO
SKIN ABOUT EVERY CM.

PICTORIAL

TIP: HOLD SKIN UP & AWAY FROM
CATHETER WHEN APPLYING
WOUND CLIP TO AVOID
PUNCTURE OF THE CATHETER.

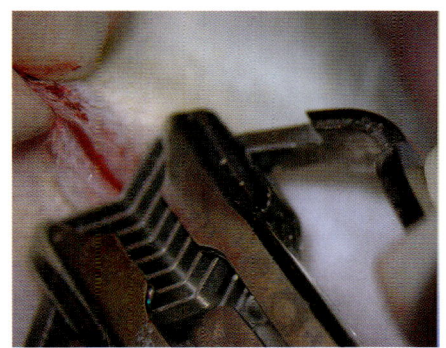

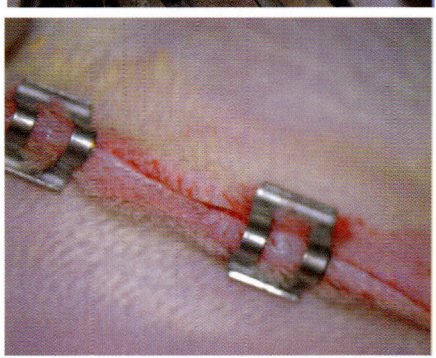

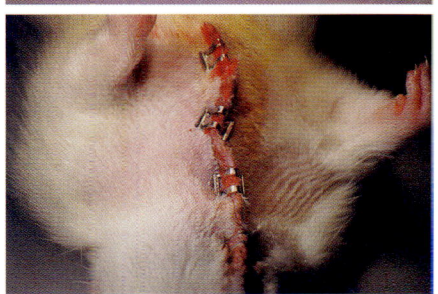

MATERIALS	PROCEDURE	PICTORIAL
Vetbond	Use one drop of vetbond in between wound clips & press edges of skin together to seal wound.	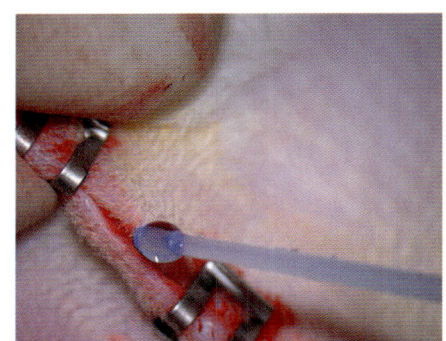
Marker	Flip over rat to sternal recumbent position. Use marker to identify jugular vein catheter @ end.	
Hemostat	With scissors action, create large pocket posterior to hole.	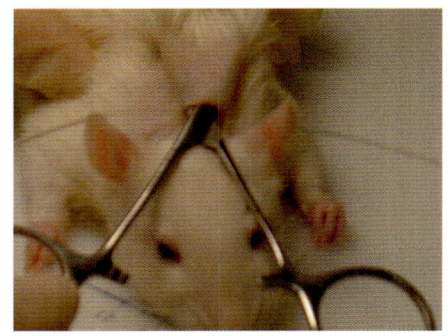

MATERIALS

PROCEDURE

PICTORIAL

TIP: DO NOT BEND CATHETER WHILE COILING. IT WILL PREVENT ANY SUBSEQUENT BLOOD FLOW.

COIL EXCESS CATHETER LENGTHS SEPARATELY & TUCK INTO POCKET.

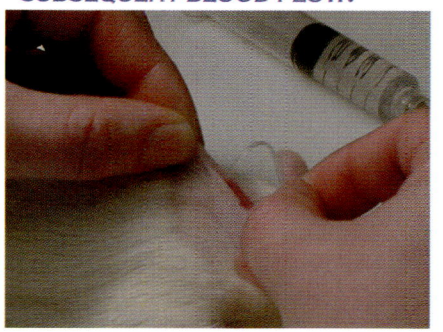

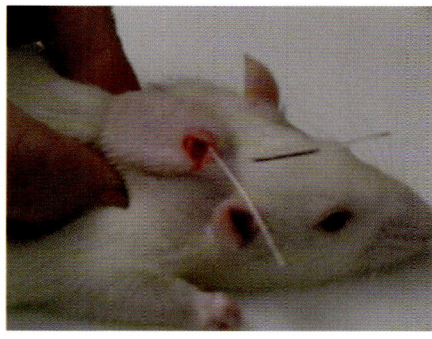

WOUND CLIPS
WOUND CLIP APPLIER

LIFT SKIN & TWO CATHETERS (IN OPPOSITE DIRECTIONS) & SECURE WITH ONE WOUND CLIP.

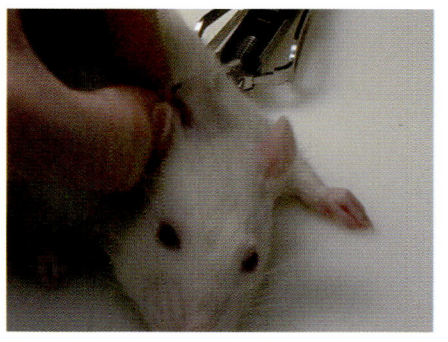

63

MATERIALS

WOUND CLIPS
WOUND CLIP APPLIER

PROCEDURE

DO NOT ALLOW WOUND
CLIP TO PUNCTURE
CATHETER INSIDE
POCKET.

TIP: FOR STUDIES
LONGER THAN 24
HOURS, ONE MAY TUCK
CAROTID CATHETER END
BACK THROUGH CLIP
FOR ADDED SECURITY.

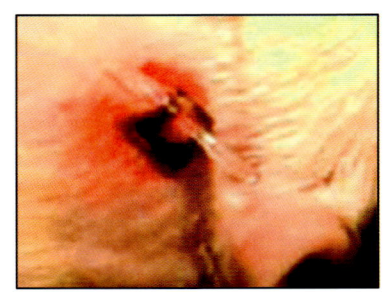

PICTORIAL

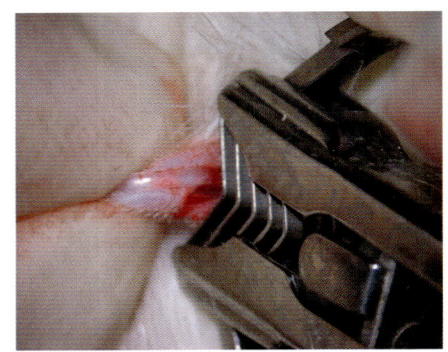

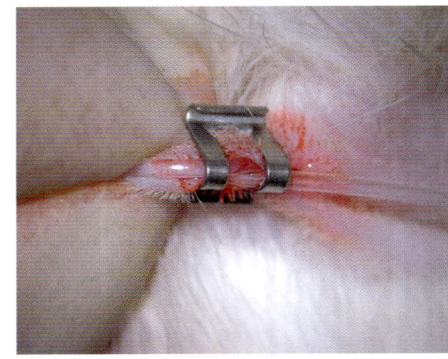

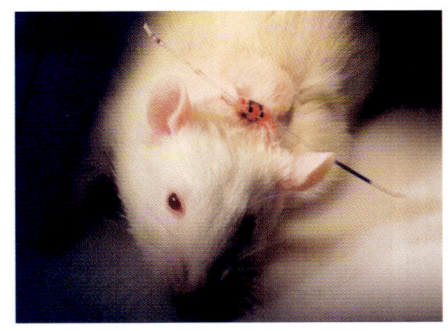

MATERIALS

0.1 MG/ML ATROPINE
SULPHATE

HEATING PAD

TRANSGEL®
RODENT CHOW
RODENT CAGE

PROCEDURE

ADMINISTER IP
INJECTION OF 0.3 ML
ATROPINE SULPHATE TO
BLOCK VAGAL
STIMULATION. PLACE
ANESTHETIZED RAT ON
SIDE ON HEATING PAD
(LOW) UNTIL ABLE TO
RIGHT SELF. CHANGE
SIDES EVERY 15 – 30
MINUTES. PLACE RAT IN
CAGE WITH TRANSGEL &
FOOD ON BOTTOM OF
CAGE. SINGLY HOUSE
RATS TO PREVENT
CHEWING OF EACH
OTHERS' CATHETERS.
ALLOW MINIMUM OF 17-
24 HOURS RECOVERY
BEFORE USE IN STUDY.

PICTORIAL

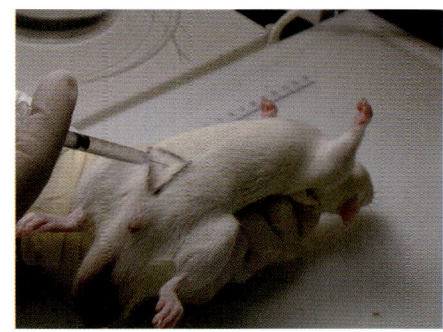

NOTE: POST-OPERATIVE PAIN
MANAGEMENT IS RECOMMENDED
UNLESS THERE IS SCIENTIFIC
EVIDENCE THAT IT WILL
INTERFERE WITH STUDIES SUCH
AS PHARMACOKINETICS AND
CNS MODELS.

BLOOD DRAWING VIA CAROTID CATHETER

It is generally agreed among experts that 10% or less of the circulating blood volume can be removed without resulting gross abnormalities. With this amount of blood loss, baroreceptor-initiated reflexes can cause a release of cholinergics from the medulla and sympathetic nerve endings. Increased heart rate results from constricted arteriole beds and venous reservoirs in the muscle and skin. There is a minimal effect on blood pressure and cardiac output. The slow rate of blood volume replacement is caused by a secretion of antidiuretic hormone and activation of the renin-angiotensin system (ILAR, 1989). Blood volume of a rat is 58 – 70 ml/kg (Diehl, 2001). If we base our calculations on the midpoint of the range (64 ml/kg), a rat weighing between 225 – 320g would have a blood volume of 14.4 – 20.5 mL; therefore 1.4 – 2.1 mL can be withdrawn safely. The formula for chronic, repeated blood sampling is 0.6 ml/kg/day (Joint Working Group, *Lab Animals*, 1993).

Cardiac output and blood pressure are depressed at 15 – 20% blood loss. A massive cholinergic release leads to tachycardia and compensatory tachypnea. Extreme arteriole constriction redistributes blood away from the gut and skin. It is reported that 20% blood loss can cause a 25% reduction in arterial pressure (Ploucha, 1986). Lower oxygen levels produce anaerobic glycolysis. Increased plasma lactate produces metabolic acidosis. Venous constriction helps to maintain venous return. Finally, interstitial fluid is transported to intravascular compartments; helping to slowly replace fluid volume. In humans, this vasovagal response produces clinical signs of nausea, dizziness, and blurred vision (ILAR, 1989).

If one withdraws 30 – 40%, hemorrhagic shock is possible. In addition to tissue anoxia, hypercapnia, and acidosis, a poorly perfused pancreas depresses myocardial function. This cascades into cell injury, irreversible tissue damage, organ failure, and death. Poor blood flow in the medulla reduces any compensatory reflexes (ILAR, 1989). Hematocrit values may not be accurate in the 72 hours following blood loss due to a proportional RBC: plasma reduction (Wintrobe, 1981). More than 40% total blood removal can result in mortality of half the number of rats used.

Blood volumes should not be removed too frequently or too rapidly. Even small volumes in this manner can lead to acute shock or anemia. For required intensive blood sampling in 8 – 24 hour pharmacokinetic studies, it is acceptable to remove up to 20%. The rat should be sacrificed at the terminal bleed. It is reported that extreme loss of blood can induce feelings of anxiety and disorientation and it is considered inhumane and unethical in the treatment of laboratory animals (AVMA Panel, 2000). Fluid replacement is acceptable with saline, ideally warmed at 30 - 35°C.

Symptoms of anemia include pale color (eyes, ears, tongue, and extremities), decreased activity, and increased respiration rate when active. Symptoms of shock include dry membranes, low body temperature, restlessness, and hyperventilation.

Troubleshooting blood drawing

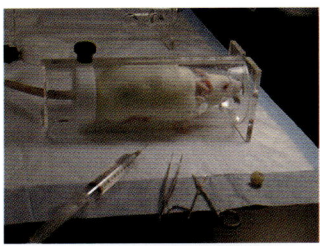
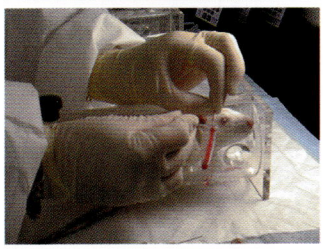

Several problems can be encountered when withdrawing blood from the carotid catheter. The following recommendations should assist you in obtaining blood samples with the least difficulty after catheterization. The rat should have a minimum of 17 – 24 hours to recover from surgery. Carefully restrain the rat in a broome restrainer or any restrainer with access to the catheter. The catheter can be pulled through the opening. Using your thumbnail or a hemostat, occlude the catheter while cutting off the plugged end with scissors. Release the hemostat or your thumbnail to collect free flowing blood. Flush back with heparinized saline (100 IU/ml saline) using 22-gauge needle and plug with obturator. Tuck excess catheter back into pocket after removing rat from restrainer.

1) Make sure the rat is relaxed and acclimated to restrainer. Stress can cause vasoconstriction.
2) A small clot could form at the implanted end of the catheter. Often this is a flap that allows injection but not withdrawal. One may try to gently withdraw first. Too much force will collapse the vessel. Next try short, quick bursts of injected heparinized saline to dislodge.
3) If the catheter was tucked in too far after surgery, it is possible that the exposed end will disappear into the pocket. Often the body temperature will cause the bone wax to melt and the rat will bleed to death. This is evidenced by a large hematoma at the back of the neck. This can be avoided by using solid obturators.

4) Individual housing will keep fellow rats from chewing each other's catheters. However, if the catheter is left out too far, the rat can reach it and chew by itself. The rat will bleed to death and blood is generally shaken all over the cage.

5) A rat left on the heating pad too long after surgery without proper supervision can cause the bone wax to melt and bleeding to occur. Again, solid obturators may be chosen to avoid this condition.

6) Make sure the catheter is not too tangled or crimped inside the pocket. If no blood flows and it is impossible to inject saline, this may be the case. You can try to pull out a generous length of catheter to check and tuck back in properly.

7) Due to increased blood pressure following surgery, there may be a small amount of blood in the implanted catheter end. Withdrawing gently with a syringe can dislodge clotted blood that has formed.

8) Do not over flush with heparinized saline. It can cause internal bleeding.

9) The catheter end can twist slightly at the point of insertion into the artery once the rat is recovered and in its normal position. Leaning against the vessel wall can block the implanted catheter end. It is possible to put your finger or a forcep into the restrainer and adjust the angled position of the neck by tipping the chin upward.

10) One may try to palpate the neck gently to regain blood flow.

11) To dislodge clots, one may insert a small wire into the catheter followed by gentle withdrawal by syringe to extract dislodged clot.

12) To maintain patency over several days, flush catheter daily.

Recommended Administration Volumes

Dose in feed	5 g/100 g/day
Dose in water	10 ml/100 g/day
Gavage	5 – 10 ml/kg
Intravenous	1 – 1.5 ml/kg
Continuous IV	1 ml/kg/hr
Intraperitoneal	10 – 30 ml/kg
Subcutaneous	10 – 30 ml/kg
Intramuscular	1 ml/kg

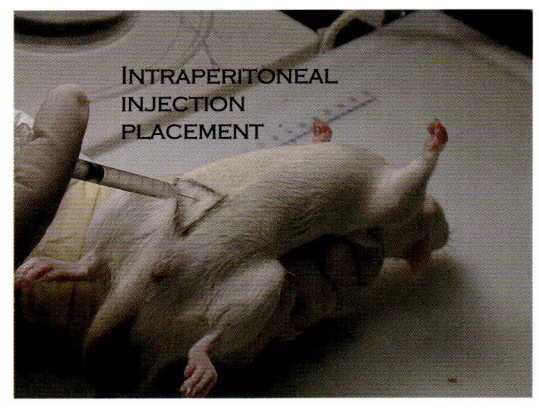

INTRAPERITONEAL INJECTION PLACEMENT

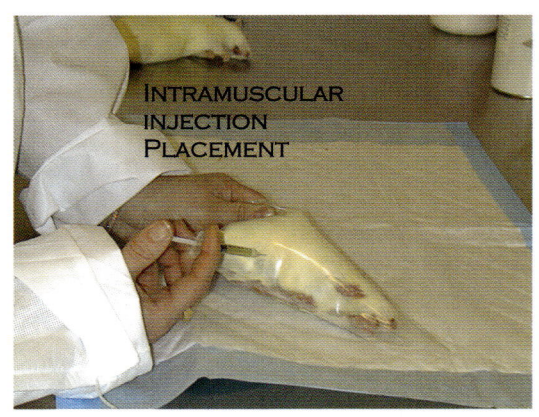

INTRAMUSCULAR INJECTION PLACEMENT

Use of venous catherization

The primary use of venous catheterization is for intravenous administration. This route is advantageous due to the predictability and reproducibility of drug kinetics. One is able to deliver agents in a controlled manner via pumps. Intravenous deliveries bypass the stomach and, therefore, avoid gastric pH and first pass metabolism issues. There is rapid onset at the target site. This is the route of choice for protein and peptide-based drugs.

There are several considerations to note prior to intravenous delivery. One must be aware of possible vascular irritation and damage caused by the test substance, leakage out of the vessel, volume overload (hemodilution), hemolysis, and infection. One must also carefully consider the solubility and pH of the test substance. Solution should be 7.4 pH adjusted and completely soluble. Any particulates will cause an emboli.

Methods in vascular infusion biotechnology in research with rodents by Dr. Nolan (2002) is an excellent resource for this application.

Intravenous injections may be given as:
- **Bolus** – equal to or less than 20 seconds, may be given manually using a 22 gauge needle to connect to the jugular catheter
- **Slow push** – usually between 1 – 5 minutes, may be given manually or via pump
- **Continuous infusion** – usually hours to days to reach a steady state in the plasma, must be delivered via pump. The rat is fitted into a harness and the catheters threaded through a

protective sheath to an overhead swivel and infusion pump. Alternatively, an osmotic pump filled with test material can be implanted SC and attached to the catheter.

<u>Vendors of continuous infusion equipment</u>:
- ➢ Instech Solomon
  5209 Militia Hill Road
  Plymouth Meeting, PA 19462
  800-443-4227
  www.instechlabs.com

- ➢ Lomir Biomedical, Inc.
  99 East Main Street
  Malone, New York 12953
  518-483-7697
  www.lomir.com

- ➢ Bioanalytical Systems, Inc. (BAS)
  2701 Kent Avenue
  West Lafayette, Indiana 47906
  www.bioanalytical.com

Use of arterial catheterization

The most common use for arterial catheterization is serial blood sampling. Catheterization of the carotid artery provides a sampling procedure, which has technical ease and greatly reduces stress to the rat. Other uses include CNS delivery and cardiac monitoring.

➢ **Blood sampling** – Serial sampling may be performed manually or via robotic samplers. Automatic blood samplers (ABS) provide many benefits including fluid replacement, less supervision, precise timepoints, and collection of inconvenient timepoints. Consider preparatory time, maintenance and cost required when choosing between manual and ABS methods. A combination of the two can be quite efficient and serve many purposes. Remember sampling volume criteria to avoid stress to the rat. *Methods in vascular infusion biotechnology in research with rodents* by Dr. Nolan (2002) is an excellent resource for this application.

➢ **Intra-arterial administration** – Reversing direction of the catheter insertion to a rostral placement, allows for delivery of test substances to the brain. Research in CNS indications, such as stroke and Alzheimer's, benefit from this administration by directly reaching the brain without metabolism.

➢ **Cardiac parameter monitoring** – Carotid catheterization is useful for monitoring blood pressure, heart rate, ECG, temperature and blood gases. The catheter is attached to transducers outside or inside the body.

Vendors of ABS equipment:

➢ Instech Solomon
  5209 Militia Hill Road
  Plymouth Meeting, PA 19462
  800-443-4227
  www.instechlabs.com

➢ Bioanalytical Systems, Inc. (Culex)
  2701 Kent Avenue
  West Lafayette, Indiana 47906
  www.bioanalytical.com

➢ DiLab Inc.
  11 Goldsmith Street
  Littleton, MA 01460
  888-844-3633
  www.dilab.com

Vendors of blood pressure equipment:

➢ ADInstruments
  2205 Executive Circle
  Colorado Springs, CO 80906
  719-576-3970
  www.adinstruments.com

➢ Data Sciences International (DSI)
  4211 Lexington Avenue,
  North, Suite 2244
  St. Paul, MN 55126
  800-262-9687
  www.datasci.com

- Mini Mitter
  20300 Empire Avenue, Bldg B-3
  Bend, OR 97701
  800-685-2999
  www.minimitter.com

RAT BIODATA

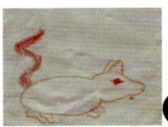

General

[1]Adult Weight (8 - 10 weeks old)
Male Sprague Dawley	275 g
Female Sprague Dawley	225 g
[1]Life Span	2 – 5 years
[1]Body Surface Area	0.03 – 0.06 cm^2
[1]Chromosome Number (Diploid)	42
[1]Water Consumption	80 – 110 ml/kg/day
[1]Food Consumption	100 g/kg/day
[1]Body Temperature	37.5°C (99.5°F)
Floor Area (Sapanski, 1984)	29 – 40 in^2

[9]**Anatomy:** Harderian gland secretes red pigmented tears, color-blind, lacking some water taste receptors, no tonsils/ gall bladder/ sweat glands, thermoregulates via tail

[9]**Behavior:** Nocturnal, passive, curious, intelligent, practice coprophagy (recycle feces), light and noise sensitive, group or singly housed

[7]Organ Weights
Adrenals	0.05 g
Brain	1.8 g
Heart	1.0 g
Intestine	11.25 g
Kidneys	2.0 g
Liver	10.0 g
Lung	1.5 g
Spleen	0.75 g
[9]Total Body Water	167 ml*
[9]Intracellular Fluid	92.8 ml*
[9]Extracellular Fluid	74.2 ml*
Growth Rate	25 g/week
	*250 g rat

Reproductive

[1]Breeding Age	50 – 60 days
[1]Breeding Season	none
Breeding Life (Sapanski, 1984)	1.5 years
[1]Gestation	21 – 23 days
[10]Pseudopregnancy Length	12 days
[10]Duration of Single Cycle	4 - 5 days

[10]Stages of Sex Cycle
Polyestrous Female
Stage 1 dioestrous	6 hours
Stage 2 pro-estrous (early)	60 hours
Stage 3 pro-estrous (late)	12 hours
Stage 4 oestrous	10 -20 hours
Stage 5 metoestrous	8 hours
[5]Time of Ovulation	8 – 11 hours *after estrous*
[5]Type of Ovulation	spontaneous

[5]Mating Methods	Monogamous Or trios	Muscle	7.5
[5]Fertilization	7 – 10 hours *after ovulation*	Portal Vein	9.8
		Skin	5.8
[5]Number of Eggs Shed	10 +	Spleen	0.63
[5]Viability of Eggs	10 -12 hours		
[5]Litter Frequency	7 – 9 /year	[1]Heart rate	330 - 480 beats/min
[9]Mammary Glands	6 pairs	[1]Blood Pressure – systolic	88 – 184 mmHg
[9]Litter Number	8 – 14 pups	[1]Blood Pressure – diastolic	54 – 145 mmHg
[9]Birth Weight	5 – 6 g	[1]Cardiac Output	10 – 80 ml/min
[9]Eyes Open	10 – 12 days	[1]Minute Volume	0.05 – 0.101 ml
[9]Weaning Age	21 days	[1]Stroke Volume	1.3 – 2.0 ml/beat
		[1]Plasma pH	7.4
		[1]Plasma CO_2	22.5 mM/l
		[1]Plasma CO_2 Pressure	40 mmHg

Cardiovascular

[7]Blood Volume	58 ml/kg
[7]Hematocrit	46%
[7]Blood pH	7.38
[7]Plasma Volume	31.3 ml/kg
[7]Plasma Albumin	31.6 mg/ml
[7]Plasma α-1-ACG	18.1 mg/ml

[7]Blood Flow	(ml/min)
Adipose	0.4
Brain	1.3
Heart	3.9
Hepatic Artery	2.0
Intestine	7.5
Kidneys	9.2
Liver	13.8

[1]Leukocyte Counts

Total	14 x 10^3/ml
Neutrophils	22%
Lymphocytes	73%
Monocytes	2.3%
Eosinophils	2.2%
Basophils	0.5%

[1]Platelets	1240 x 10^3/ml
[1]Packed Cell Volume (PCV)	46%
[1]Red Blood Cells	7.2 – 9.6 x 10^6/min3
[1]Hemoglobin	15.6 g/dl
[6]Clotting Time	20 seconds
[10]Osmolality	321 mmol/kg
[10]Whole Blood Specific Gravity	1.05
[9]Mean Corpuscular Volume (MCV)	
	46 – 65 fl

[9]Mean Corpuscular Hb Concentration (MCHC)
 31 – 40 g/dl
[9]Mean Corpuscular Hb (MCH) 18 – 23 pg
[9]Reticulocytes 0 – 25%

Metabolism

[5]Basal Metabolic Rate of 300g rat (surface area of 0.04m2 and 3,357,813 joules/m^2/day = 802 kcal/m^2/day)

[3]B-Glucuronidase Activity
Proximal Small Intestine = 304 nmol substrate/hr/g
Distal Small Intestine = 1341 nmol substrate/hr/g

[7]Liver Cytochrome P450 = 0.98 nmol/mg protein

Respiratory

[9]Respiratory Rate	70 – 115 breaths/ min
[3]Oxygen Consumption	0.84 ml/hr/g
[5]Ventilation Rate	66 – 210 breath/min
[5]Tidal Volume	0.60 – 1.25 cm^3/g/hr
[9]Trachea diameter	1.6 – 7.7 mm
[9]Total lung capacity	9.9 – 12.7 ml

Urine

[1]pH	7.3 – 8.5
[1]Specific Gravity	1.04 – 1.07
[3]Urine Flow	50 ml/day
[2]Osmolality	2442 mOsm/kg
[2]Protein	<30 mg/dl
[2]Creatinine	5.5 mg/100g/day
[2]Potassium	2.2 mEq/100g/day
[2]17-ketosteroid	16.4 µg/100g/day
[10]Urea	442.5 mmol/l
[10]Ureate	1.7 mmol/l
[10]Na+	229 mmol/l
[10]K+	149.5 mmol/l
[10]Ca^{2+}	0.7 mmol/l
[7]Number of Glomeruli	2.9 x 105/kg

[9]L-amino acid oxidase in kidneys is significant only in rats

Gastrointestinal

[7]Small Intestine Transit Time	88 min
[7]Small Intestine Length	0.1 – 0.15 m
[7]Large Intestine Length	0.02 – 0.03 m
[7]Whole Intestine Volume	11.25 ml
[7]Gut Lumen Volume	8.8 ml
[7]Fasted Stomach pH	3.0 – 3.8
[7]Postprandial Stomach pH	2.3 – 4.5

[7]Fasted Intestine pH 6.9 – 7.8
[7]Feces pH 6.9
[9]Duodenum length 10 cm
[9]Jejunum length 100 cm
[9]Ileum length 3 cm

 **Bile**

[7]Bile Flow 90 ml/kg/day

 **Brain**

[7]Blood Volume in Brain 11 µl/g tissue
[7]CSF Flow 2.2 µl/min
[9]CSF Volume 234 - 266 µl
[4]Choroid Plexus Blood Volume 13.1 µl/g tissue

[8]Solute Concentrations in CSF	mEq/Kg H_2O
Na	152
K	3.36
Ca	2.22
Mg	1.77
Glucose	5.38
Pyruvate	0.184
Lactate	2.08
[9]CSF pressure	34 – 42 mmHg

Clinical Chemistry Values

[10]Total proteins	63 g/l
[10]Albumin	28 g/l
[10]α-1 Globulins	4.6 g/l
[10]α-2 Globulins	3.5 g/l
[10]β-Globulins	5 g/l
[10]Gamma Globulins	4.4 g/l
[10]Urea	6.9 mmol/l
[10]Ureate	0.6 mmol/l
[10]Glucose	10.1 mmol/l
[10]Creatinine	42.5 µmol/l
[10]Creatine Clearance	1.2 ml/min
[10]Total Lipids	2.3 g/l
[10]Phospholipid	0.05 g/l
[10]Cholesterol	1.9 mmol/l
[10]Neutral Fat	0.8 g/l
[10]Bilirubin	2 µmol/l
[10]Aspartate aminotransferase	82 i.u./l
[10]Creatinine kinase	368 i.u./l
[10]Gamma-glutamyl transpeptidase	10 i.u./l
[10]α-hydroxybutyrate dehydrogenase	71 i.u./l
[10]Iron Binding Capacity	101 µmol/l
[10]Specific Gravity (cm water)	1.21

[10]Fe^{3+}	28 µmol/l
[10]Na^+	135 mmol/l
[10]K^+	4.9 mmol/l
[10]Ca^{2+}	2.6 mmol/l
[10]Cu^+	17.8 µmol/l
[10]Mg^{2+}	1.3 mmol/l
[10]Cl^-	100 mmol/l

---

[1]Cassella J. *et al.*, *The Rat Nervous System*, John Wiley and Sons, New York, 1997.
[2]Baker H. *et al.* (eds.), *The Laboratory Rat, Volume 1: Biology and Diseases*, Academic Press, New York, 1979.
[3]Davies B. and Morris T., Physiological parameters in laboratory animals and humans, *Pharmaceutical Research*, **10**(7): 1093 – 1095, 1993.
[4]Davson H. and Segal M. (eds.), *Physiology of the CSF and Blood-brain Barriers*, CRC Press, New York, 1996.
[5]Inglis J., *Introduction to Laboratory Animal Science and Technology*, Pergamon Press, New York, 1980.
[6]Joint Working Group on Refinement, Removal of blood from laboratory mammals and birds, *Laboratory Animal*, **27**: 1 – 22, 1993.
[7]Kwon Y., Handbook of Essential Pharmacokinetics, Pharmacodynamics, and Drug Metabolism for Industrial Scientists, Kluwer Academic Press, New York, 2001.
[8]Ohno K. *et al.*, Lower limits of cerebrovascular permeability to nonelectrolytes in the conscious rat, *American Journal of Physiology*, **235**(3): H299 – H307, 1978.
[9]Sharp P. and LaRegina M., *The Laboratory Rat*, CRC Press, New York, 1998.
[10]Waynforth H. and Flecknell P, *Experimental and Surgical Technique in the Rat*, Academic Press, New York, 1992.

Spatial Definitions

The author has attempted to relate the step-by-step procedure in a clear, universally understood language. However, in the world of surgery and medicine, the terminology used is a language of its own. This reference section should give the reader a brief idea of the most common terms; their definition and usage.

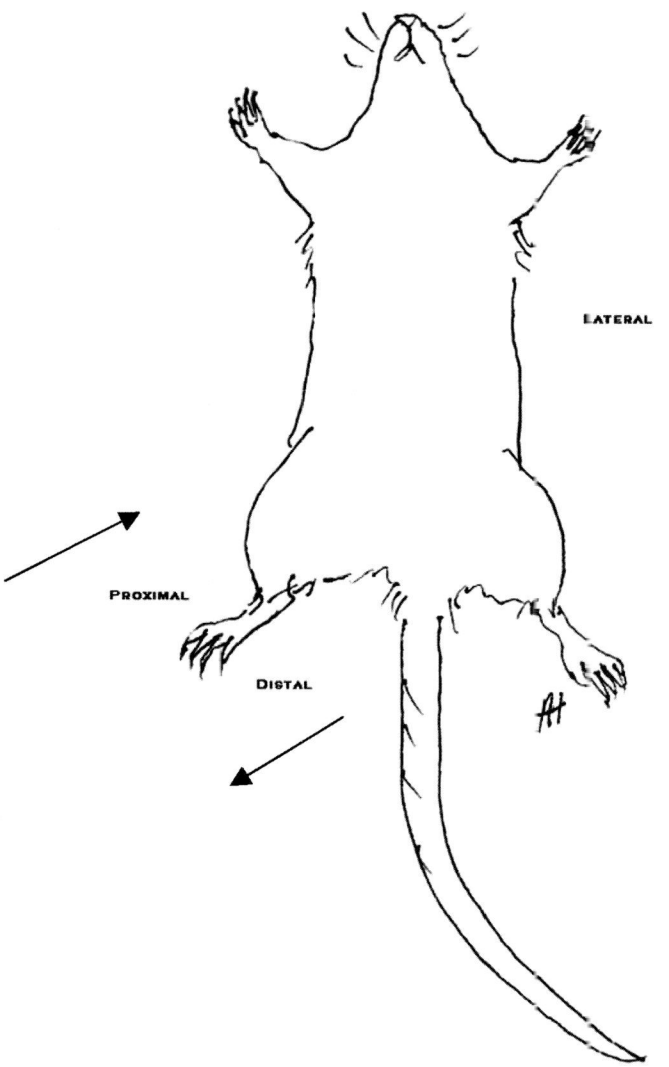

LATERAL

PROXIMAL

DISTAL

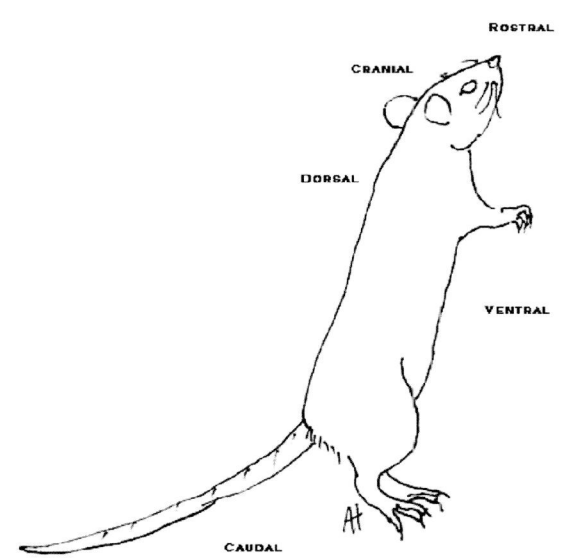

ROSTRAL

CRANIAL

DORSAL

VENTRAL

CAUDAL

<u>Sagittal plane</u>: longitudinal cross-section from cranial to caudal
<u>Coronal plane</u>: longitudinal cross-section from dorsal to ventral
<u>Transverse plane</u>: lateral cross-section at midline
<u>Median plane</u>: longitudinal cross-section at the midline

An example of correct usage is as follows:

"…an incision was made in the ventrolateral aspect of the neck. The right external jugular vein was dissected free of surrounding fascia and stabilized with two loops of 4-0 silk suture. The distal loop was tied to ligate the vein on the cranial aspect. Venotomy was made approximately five millimeters cranial to the site of crossover by the pectoralis major muscle… A subcutaneous tunnel was created from the neck area to the dorsum, and the catheter was pulled through to exit through an incision in the interscapular area. A subcutaneous pocket in the interscapular region was created by blunt dissection." ~ Foley, 2002.

LATIN TERMS

Abbreviation	Expanded form	Definition
ad lib.	ad libitum	at pleasure
A.M.	ante meridian	before noon
b.	bis	twice
b.i.d.	bis in die	twice a day
bol.	bolus	a large pill
brevis	brevis	short
cap.	capiat	let the patient take
caps.	capsula	a capsule
i.c.	inter cibos	between meals
lb.	libra	pound
m.	mane	in the morning
m. dict.	more dicto	as directed
m.t.d.	mitte tales doses	send such doses
n.	naris	nostril
nebul.	nebula	a spray
n. et m.	nocte maneque	night and morning
noct.	nocte	at night
o.d.	octus dexter	right eye
omn. Hor.	omni hora	at every hour

Abbreviation	Expanded form	Definition
o.s.	oculus sinister	left eye
p.c.	post cibos	after meals
P.M.	post meridiem	after noon
p.o.	per os	by mouth
ppt.	praecipitus	precipitated
pro rect.	pro recto	rectal
q., qq.	quodque,quaeque	each, every
q.i.d.	quarter in die	four times a day
qq. hor	quaque hora	every hour
q.s.	quantum sufficiat	a sufficient quantity
quot. Op. sit	quoties opus sit	as often as necessary
sol.	solubilis	soluble
s.o.s.	si opus sit	if there is need
ss.	semis	one half
stat.	statim	immediately
syr.	syrupus	syrup
tab.	tabella	tablet
t.i.d.	ter in die	three times a day
ung., ungt.	unguentum	ointment
vesp.	vesper	evening

RESOURCES

Vendors

➤ **Braintree Scientific**
P.O. Box 850929
Braintree, Massachusetts
02185
www.braintreesci.com
781-843-2202
Decapicones, gas chamber, tubing, tethers, suture, sterilizer, balance, wound clip system, warming pad, heat lamp, restrainer, microscope, PVC Rat, ThermoCare ICU

➤ **Ancare**
P.O. Box 814
Bellmore, New York
11710
www.ancare.com
919-620-7504
Nestlets, caging

➤ **Henry Schein**
135 Duryea Road
Melville, New York
11747
www.henryschein.com

800-V-SCHEIN
Sterilizer bags, needles, heparin,vetbond,atropine sulphate, IV saline bags, paralube, sterile saline, buprenorphine, ketamine, acepromazine, xylazine, gauze, betadine wipes, alcohol wipes, antibiotic

➤ **Samuel Perkins Co., Inc.**
497 Beale Street
Quincy, Massachusetts
02169
617-773-3600
Heating pad, bone wax

➤ **George Tiemann**
25 Plant Avenue
Hauppauge, New York
11788
www.georgetiemann.com
800-843-6266
3-0 silk suture (100 yards)

- **Harvard Apparatus**
  22 Pleasant Street
  South Natick, Massachusetts
  01760
  800-272-2775
  Glass bead sterilizer, spear-shaped sponges,
  ThermoCare ICU

- **Roboz**
  PO Box 10710
  Gaithersburg, Maryland
  20898
  www.roboz.com
  301-590-0055
  Surgical instruments, wound clips and applier,
  wound clip remover

- **Leica**
  90 Boroline Road
  Allendale, NJ
  07401
  www.leica-microsystems.com
  800 526 0355
  MZ6 dissecting scope

- **Baxter Scientific**
  One Baxter Parkway
  Deerfield, IL
  60025

www.baxter.com
800-422-9837
Surgical board

- **J.A. Webster**
  86 Leominster Road
  Sterling, Massachusetts
  01564
  800-225-7911
  A5 clippers, 40mm clipper blade, fenestrated
  towels, 5% dextrose

- **VWR**
  1310 Goshen Parkway
  West Chester, PA 19380
  www.vwr.com
  800-932-5000
  Small and large underpads, Sartorius animal
  weighing scale and printer, gavage needles,
  broome restrainers, PE50 tubing, syringes

- **Fisher Scientific**
  2000 Park Lane Drive
  Pittsburgh, PA 15275
  www.fisherscientific.com
  800-766-7000
  Cotton applicators, insulin syringes

- **Hazard Technologies**
  406 Headquarters Drive #5
  Millersville, Maryland
  21108
  410-987-7833
  Vacuum/clipper system

- **Lab Products**
  742 Sussex Avenue
  Seaford, Delaware
  19973
  800-526-0469
  Caging equipment

- **Bio-Serv**
  One 8th Street, Suite 1
  Frenchtown, NJ
  08825
  908-996-2155
  www.bio-serv.com
  Nutra-Gel, enrichment

- **Boston Tec**
  2700 James Savage Road
  Midland, MI
  48642
  989-496-9510
  www.bostontec.com
  Hydraulic tables

- **Surgical Specialties**
  10 Dennis Drive
  Reading, Pennsylvania
  19606
  www.sharpoint.com
  800-523-3332
  Sharpoint PracticePak

- **Charles River Labs**
  251 Ballardvale Street
  Wilmington, Massachusetts
  01887
  800-LAB-RATS
  www.criver.com
  Rats, Transgel®

- **Harlan**
  P.O. Box 29176
  Indianapolis, Indiana
  46229
  317-894-7521
  Rats

- **Clear H2O**
  117 Preble Street
  Portland,ME
  04101
  888-493-7645
  www.clearh2o.com
  HydroGel™

- *Taconic*
  273 Hover Avenue
  Germantown, New York
  12526
  www.taconic.com
  518-537-5200
  Rats

Websites

Animal Science Associations

American Association for Accreditation of Laboratory Animal Science (AALAC)
www.aaalac.org

American Association for Laboratory Animal Science (AALAS)
www.aalas.org

American Committee on Laboratory Animal Diseases (ACLAD)
www4.ncsu.edu/unity/users/b/bweigler/Web/ACLAD/Index.html

Institute of Laboratory Animal Resources (ILAR)
www2.nas.edu/ilarhome

Lab Animal Magazine
www.labanimals.com

USDA Animal and Plant Health Inspection Service
www.aphis.usda.gov/vs/vshome.html

Food and Drug Administration (FDA)
www.fda.gov/fdahomepage.html

Laboratory Animal Science Association (LASA)
www.mandm.ncl.ac.uk/lasa.html

Laboratory Animal Welfare Training Exchange (LAWTE)
netvet.wustl.edu/org/lawte/homepg.htm

Wound closure
http://ethicon.com/page/pdf/WoundClosureManual101702.pdf

Suture Tying
www.jnjgateway.com/public/USENG/5256ETHICON_Encyclopedia_of_Knots.pdf

Disinfectants
www.fda.gov/cdrh/ode/germlab.html

Principles of Surgery
http://cal.vet.upenn.edu/surgery/index.htm

Surgical Equipment and products
www.MyNeurolab.com

Fluid Administration
Alza Scientific Products
www.alza.com

GLOSSARY

Acclimate – to become familiar with new surroundings

Amnesia – loss of memory

Anaerobic glycolysis – breakdown of sugar in absence of oxygen

Analgesia – agent which blocks the perception of pain without loss of consciousness

Anemia – condition caused by low red blood cell count

Anesthetic – agent which reduces or eliminates sensory and motor responses accompanied by loss of consciousness

Anterior – situated toward the front or head

Antibiotic – agent that kills bacteria

Anti-emetic – agent which suppresses nausea; useful for motion sickness

Asepsis – a state of sanitation in which microorganisms are greatly reduced

Autoclave – machine that sterilizes by high temperature and high pressure

Barbiturates – group of sedative drugs derived from barbituric acid

Baroreceptor – receptor that senses or controls pressure changes

Biocompatible – mixture of two substances resulting in no adverse reaction

Bronchodilation – opening of bronchi; two tubes at the lower end of the trachea

Cadaver – a dead body

Catheter – tubing through which blood or other fluids can flow

Cardiac output – the rate at which blood is pumped out of the ventricles of the heart

Catalepsy – sustained immobility

Caudal – situated towards the tail

Cautery – device used to apply direct heat for the purpose of coagulating blood

Cerebral – pertaining to the two cerebrum halves of the brain, which control motor and sensory function

Cholinergic – refers to parasympathetic nerves which release acetylcholine

Collateral – secondary

Concentric – having a common middle

Connective tissue – tissue which connects or supports other cells or tissues

Coprophagy – a nutritive practice by rodents and rabbits of recycling feces

Cornea – a convex membrane covering the iris of the eye

Cranial – situated towards the head

Cryosurgery – surgery using freezing probe in place of scalpel knife

Cumulative dosing – additive effect of multiple dosing

Decapicone – plastic cone-shaped bag used as a restraint for rodents

Diarrhea – loose stool

Dilate – to widen or enlarge

Dissociative anesthetic – an anesthetic agent which disconnects motor and sensory responses

Distal – situated away from a specified point

Diuresis – increased urine output

Dorsal recumbency – lying on the back

Dyspnea – labored breathing

Edema – an abnormal swelling due to accumulation of interstitial fluid

Embolism/emboli – blockage by solid particle, clot, or air bubble in circulation

Endothelium – membranous lining

Enzootic – a type of disease which occurs in one or more classifications of a population e.g. newborns

Extravasation – fluid escaping outside of vessel

First pass metabolism – refers to process following oral administration whereby first pass through GI and liver results in breakdown or extraction of substance and less compound reaches circulation

Gavage – a method of oral administration directly to the stomach

Gross – visible by eye only

Hematocrit – percent of packed cell volume (PCV); volume of red blood cells

Hematoma – abnormal accumulation of blood

Hemorrhagic shock – a circulatory imbalance due to extreme loss of blood volume

Hepatic – pertaining to the liver

Homeostasis – internal equilibrium maintained by adjusting physiological processes

Hybrid – mixture of characteristics from two or more sources

Hypercapnia – excessive CO_2 in blood

Hyperglycemia – a condition characterized by excessive sugar in the bloodstream

Hyperoxic ventilation – respiration resulting in increased oxygen

Hyperventilation – an increase in respiratory rate, depth, and duration

Hypothermia – below normal body temperature

Hypovolemic shock – a condition characterized by a severe decrease in circulating blood volume

Hypoxia – a condition characterized by a decrease of oxygen in the tissues

Interstitial fluid – fluid between cells

Intramuscular – within a muscle

Intraperitoneal – within the abdominal cavity but not inside the abdominal organs

Intrathecal – within the subarachnoid spaces that contain cerebrospinal fluid

Intravascular – within the blood vessel

Intravenous – within the vein

Isothermic – temperature that is maintained at a constant setting

Laryngeal – of the larynx; the voice organ

Latent – dormant or inactive state

Lateral – situated away from the middle of the body

Limbic – periphery or edge of main structure

Lipid soluble – ability to dissolve in fatty acid

Medial – situated towards the middle of the body

Medulla – upper portion of spinal cord

Metabolic – intracellular process that breaks down compounds

Metabolic acidosis – depletion of alkali reserves

Mortality – rate of death

Mucous membrane – glands which secrete mucus

Nocturnal – exhibits most active behavior at night

Obturator – a device that causes a stoppage

Occlude – to cause a closure

Palpate – manual examination

Palpebral – pertaining to the eyelid

Parasympatholytic – that which neutralizes effect of parasympathetic stimulation

Patent – open and clear

Pedal – pertaining to the foot

Perioperative – surrounding the operation

Pharmacokinetics – the study of the way molecules behave in the body

Pharyngeal – pertaining to the pharynx; a cavity at the back of the mouth

Porphyrin – red pigment secreted by the harderian gland behind the eye that may protect from light

Posterior – situated toward the back

Quarantine – period of isolation from established population

Renin-angiotensin system – kidney enzyme-substrate action that affects blood pressure

Rostral – situated towards the nose

Secretion – fluid passed from gland to GI, blood, or exterior

Sedative – an agent that calms and lessens functional ability

Self-mutilation – A behavior or disease in which rats bite and scratch themselves

Sensory cortex – area of the brain pertaining to sensation

Splenocyte – cells found in the spleen

Sterile – free from microorganisms

Sternal recumbency – lying on front or sternum

Subcutaneous – underneath the skin

Subcuticular suture – sutured beneath the cuticle of the skin

Sympathetic – pertaining to autonomic nervous system; nerves which extend to all muscles

T cell – a type of immune cell called a lymphocyte

Tachycardia – rapid beating of heart

Tachypnea – abnormal respiration

Tissue anoxia – lack of oxygen in tissues

Thermoregulation – self-regulation of body heat

Transducer – device which transfers power between systems for the purpose of measuring pressure, temperature, speed, etc.

Unilateral – pertaining to one side

Vagal tone – refers to regulation of heart rate by manipulations to the vagus nerve

Vasoconstriction – the narrowing of a vessel wall lumen

Vasodilation – the widening of a vessel

Vasospasm – vessel wall constrictions

Vasovagal – pertaining to reflex action between vagus nerve and circulation causing heart to slow down; blood pressure and oxygen decrease result in patient fainting

Ventral – situated towards the front of a rat

Venotomy – an incision performed on a vein

Zoonotic – a type of disease which is transmitted between animals and humans

BIBLIOGRAPHY

Arlund E. and Heyl W., Low maintenance high patency vascular cannulation in the rat, *AALAS poster in Taconic Technical Library*, 1997.

Arnander C, *et al.*, Long-term stability in vivo of a thromboresistent heparinized surface, *Biomaterials*, **8**: 496 – 9, 1987.

AVMA Panel, Report of the AVMA Panel on Euthanasia, *JAVMA*, **218**(5): 669 – 696, 2000.

Baker H., *et al.* (eds.), *The Laboratory Rat, Volume 1: Biology and Diseases*, Academic Press, New York, 1979.

Bennet B., *et al.*, Essentials for research personnel, USDA, 1994. (Brown University)

Biomethodology in the Rat (website), http://research.uiowa.edu/animal/?get=rat

Bojrab M. (ed.), *Current Techniques in Small Animal Surgery*, Williams and Wilkins, Inc., Baltimore, 1998.

Borchardt R., *et al* (eds.), *Models for Assessing Drug Absorption and Metabolism*, Chapter 15: Brain Perfusion Systems for Studies of Drug Uptake and Metabolism in the Central Nervous System, pp. 285 – 307, Plenum Press, New York, 1996.

Bradfield J, *et al.*, Behavioral and physiologic effects of inapparent wound infections in rats, *Lab Animal Science*, **42**: 572 – 8, 1992.

Brown M. *et al.*, Chapter 6: Perioperative Care, in *Essentials for Animal Research: A Premier for Research Personnel*, Florida State University Website: http://www.fsu.edu/~FSULAR/esperi.html

Burt M.E., *et al.*, Chronic arterial and venous access in the unrestrained rat, *Am J Physiol*, **238** (7): H599 – 603, 1980.

Cassella J. *et al.*, *The Rat Nervous System*, John Wiley and Sons, New York, 1997.

Cocchetto D.M. and Bjornsson T.D., Methods for vascular access and collection of body fluids from the laboratory rat, *J Pharm Sci*, **75** (5): 465 – 492, 1983.

Colletti, A., Personal observation and experience.

Cooper D.M. *et al.*, The thin blue line: a review and discussion of aseptic technique and postprocedural infections in rodents, *Contemporary Topics in Laboratory Animal Science*, **39**: 27 – 32, Nov 2000.

Cox G. *et al.*, A comparison of heparinized saline irrigation solutions in a model of microvascular thrombosis, *British Journal of Plastic Surgery*, **45**: 345 – 348, 1992.

Coyle P. and Panzenbeck M., Collateral development after carotid artery occlusion in Fisher 344 rats, *Stroke*, **21**: 316 – 321, 1990.

Cruz J.I., *et al.*, Observations on the use of medetomidine/ketamine and its reversal with atipamezole for chemical restraint in the mouse, *Laboratory Animals*, **32**: 18 – 22, 1998.

Cunliffe-Beamer T., Applying Principles of Aseptic Surgery to Rodents, *AWIC Newsletter*, **4**(2): 3 – 6, 1993.

Daniel R. (ed), *Reconstructive Microsurgery, Chapter 7: Normal Blood Flow and Chapter 8: Abnormal Blood Flow*, pp 65 – 88, Little, Brown, London, 1979.

Davies B. and Morris T., Physiological parameters in laboratory animals and humans, *Pharmaceutical Research*, **10**(7): 1093 – 1095, 1993.

Davson H. and Segal M. (eds.), *Physiology of the CSF and Blood-brain Barriers*, CRC Press, New York, 1996.

Diehl K.H., *et al.*, A good practice guide to the administration of substances and removal of blood, including routes and volumes, *J Appl Toxicol*, **21**(1): 15 – 23, 2001.

Dunn R., Anesthetics in elasmobranchs: A review with emphasis on halothane-oxygen-nitrous oxide, *Journal of Aquaculture and Aquatic Science*, **5(3)**, 1985.

Fagin K.D., *et al.*, Effects of housing and chronic cannulation on plasma ACTH and corticosterone in the rat, *Am J Physiol*, **245**: E515 – 520, 1983.

Flecknell P., Assessment and Alleviation of Post-operative Pain, *AWIC Newsletter*, **8**(3 – 4), Winter 1997/1998.

Flecknell P., *Laboratory Animal Anesthesia*, Academic Press, New York, 1996.

Flynn L.A. and Guilloud R.B., Vascular catheterization: Advantages over venipuncture for multiple blood collection, *Lab Anim*, **17** (6): 29 – 35, 1988.

Foley P., *et al.*, Effect of covalently bound heparin coating on patency and biocompatibility of long-term indwelling catheters in the rat jugular vein, *Comparative Medicine*, **52**(3): 243 – 248, 2002.

Fossum T. *et al.*, *Small Animal Surgery* (2nd Edition), Mosby, St. Louis, 2002.

Fox C.E. and Beazley R.M., Chronic venous catheterization: A technique for implanting and maintaining venous catheters in rats, *J Surg Res*, **18**: 607 – 10, 1975.

Fox J., Cohen B., and Loew F. (eds.), *Laboratory Animal Medicine*, 1984.

Gentsch C., *et al.*, Different reaction patterns in individually and socially reared rats during exposures to novel environments, *Behavioural Brain Research*, **4** (1): 45 – 54, 1982.

Giner M., *et al.*, Chronic vascular access for repeated blood sampling in the unrestrained rat, *Am J Physiol*, **253** (4): H992, 1987.

Goldmann D. and Pier G., Pathogenesis of infections related to intravascular catheterization, *Clinical Microbiology Reviews*, **6**(2): 176 – 92, 1993.

Goodson W. and Hunt T., Wound collagen accumulation in obese hyperglycemic mice, *Diabetes*, **35**(4): 491 – 5, 1986.

Green C. J., *et al.*, Ketamine alone and combined with diazepam or xylazine in laboratory animals: a 10 year experience, *Laboratory Animal*, **15**(2): 163 – 170, Apr 1981.

Greene E., *Anatomy of the Rat*, Hafner, New York, 1963.

Harkin A., *et al.*, Physiological and behavioral responses to stress: what does a rat find stressful?, *Lab Anim*, **31** (4): 42 – 49, 2002.

Hayes J. and Flecknell P., A Comparison of pre- and post-surgical administration of bupivacaine or buprenorphine following laparotomy in the rat, *Laboratory Animals*, **33**, 16 – 23, 1998.

Hebel R. and Stromberg M., *Anatomy of the Laboratory Rat*, Williams and Williams, Baltimore, 1986.

Hillyer E. and Quesenberry K., *Ferrets, Rabbits, and Rodents: Clinical Medicine and Surgery*, 1997.

Hohn D., Wixon S., White W., and Benson J. (eds.), *Anesthesia and Analgesia in Laboratory Animals*, 1995.

Hoyt R. *et al.*, Introduction to Microsurgery: An emerging discipline in biomedical research, *Laboratory Animal*, **30**(9): 26 – 35, 2001.

Hoyt R. *et al.*, Microsurgical instrumentation and suture material, *Laboratory Animal*, **30**(9): 38 – 45, 2001.

Inglis J., *Introduction to Laboratory Animal Science and Technology*, Pergamon Press, New York, 1980.

Joint Working Group on Refinement, Removal of blood from laboratory mammals and birds, *Laboratory Animal*, **27**: 1 – 22, 1993.

Juli J. *et al.*, Stress measurements in mice after transportation, *Laboratory Animal*, **29**: 132 – 138, 1995.

Kasari, Aseptic Training and Techniques, in *Laboratory Animal Welfare Training Exchange* (LAWTE), 2003.

Kato R. and Kamataki T., Cytochrome P450 as a determinant of sex difference of drug metabolism in the rat, *Xenobiotica*, **12** (11):787 – 800, 1982.

Ketamine website: http://www.rxlist.com/cgi/generic3/ketamine_cp.htm

Khouri R., Avoiding Free Flap Failure, *Clinics in Plastic Surgery*, **19**(4): 773, Oct 1992.

Kwon Y., *Handbook of Essential Pharmacokinetics, Pharmacodynamics, and Drug Metabolism for Industrial Scientists*, Kluwer Academic Press, New York, 2001.

Lawson P. (ed), *Manual for Assistant Laboratory Animal Technicians*, AALAS, 1998.

Lestage P., *et al.*, A chronic arterial and venous cannulation method for freely moving rats, J *Neurosci Methods*, **13**: 223 – 229, 1985.

Ling S. and Jamali F., Effect of cannulation surgery and restraint stress on the plasma corticosterone concentration in the rat: application of n improved HPLC assay, *Journal of Pharmacology and Pharmaceutical Science*, **6**(2): 246 – 251, 2003.

Loeb W. and Quimby F. (eds.), *The Clinical Chemistry of Laboratory Animals*, Taylor and Francis, 1999.

Loget O, *et al.*, Corneal damage following continuous infusion in rats. [In: Green K, *et al.* (eds.), Advances in Ocular Toxicology, New York: Plenum Press], page 55 – 62, 1997.

Lumb W.V. and Jones E.W., *Veterinary Anesthesia*, Lea and Febiger, Philadelphia, 1973.

Luo Y.S., *et al.* (*Charles River Laboratories*), Comparison of catheter lock solutions in rats, annual AALAS meeting paper, San Diego, 2000.

McGuill M.W. and Rowan A.N., Biological effects of blood loss: Implications for sampling volumes and techniques, *ILAR News* **31** (4): 5 – 20, 1989.

Meingassner J. and Schmook F., Reference Paper, Reference Values for Crl: CD®(SD) BR Rats, 3(1): 1 – 11, *Charles River Laboratories*, 1990.

National Research Council, Chapter 3: Surgery, Chapter 4: Facilities for Aseptic Surgery, in *Guide for the Care and Use of Laboratory Animals*, National Academy Press, Washington, D.C., 1996, pp. 56 – 79.

Nolan T and Hilton K, *Methods in vascular infusion biotechnology in research with rodents*, ILAR Journal, **43** (3): 175 – 182, 2002.

Ohno K. *et al.*, Lower limits of cerebrovascular permeability to nonelectrolytes in the conscious rat, *American Journal of Physiology*, **235**(3): H299 – H307, 1978.

Phillips R, *The Heart and Circulatory System*, Carolina Biological Supply Company, Access Excellence Classic Collection, Website:
http://www.accessexcellence.org/AE/AEC/CC/heart_anatomy.html

Ploucha J.M. and Fink G.D., Hemodynamics of hemorrhage in the conscious rat and chicken, *Am J Physiol*, **251**: R846 – 850, 1986.

Richman K.A., *et al.*, Thrombocytopenia and altered platelet kinetics associated with prolonged pulmonary artery catheterization in the dog, *Anesthesiology*, **53**: 101 – 105, 1980.

Ross J., *Blood Vessels*, Website: http://greenfield.fortunecity.com/rattler/46/comparison_of_types_of_vessels.htm

Sakura S., et al., Intrathecal catheterization in the rat: improved technique for morphologic analysis of drug-induced injury, *Anesthesiology*, **85**: 1184 – 9, 1996.

Sharp P. and LaRegina M., *The Laboratory Rat*, CRC Press, New York, 1998.

Sherwood-Davis and Geck, Online *Veterinary Suture Training Manual*

Short C.E. (*ed.*), *Principles and Practices of Veterinary Anesthesia*, Williams and Wilkins, Baltimore, 1987.

Silverman J., Suckow M., and Murthy S. (*eds.*), *The IACUC Handbook*, CRC Press, Washington, D.C., 2000.

Singer A.J. and Clark, R., Cutaneous Wound Healing, *New England Journal of Medicine*, **341** (10): 738 – 746, 1999.

Slatter D. (ed.), *Textbook of Small Animal Surgery* (3rd edition), W.B. Saunders, 2002.

Smith A. and Swindle M. (eds.), *Research Animal Anesthesia, Analgesia, and Surgery,* 1994.

Steffens A.B., A method for frequent sampling of blood and continuous infusion of fluids in the rat without disturbing the animal. *Physiol and Behav*, **4**: 833 – 836, 1969.

Stewart L., *et al.*, Evaluation of postoperative analgesia in a rat model of incisional pain, *Contemporary Topics*, **42** (1): 28 – 34, 2003.

Svedsen P. and Han J., *Handbook of Laboratory Animal Science, Volume 1: Selection and Handling of Animals in Biomedical Research,* CRC Press, New York, 1994.

Tanaka S., *et al.*, T lymphopenia in genetically obese rats, *Clinical Immunology and Immunopathology*, **86**(2): 219 – 25, 1998.

Tinsley F., *et al.*, Preparation of a jugular vein catheter: use with a semiautomatic blood sampling system, *Journal of Applied Physiology*, **54**: 1422 – 6, 1983.

Tracy D., *Small Animal Surgical Nursing*, Mosby, St. Louis, 1994.

Truex R., *Strong and Elwyn's Human Neuroanatomy*, Williams and Wilkins Company, Baltimore, 1959.

Tsui B., *et al.*, A reliable technique for chronic carotid arterial catheterization in the rat, *Journal of Pharmacological Methods*, **25**(4): 343 – 352, 1991.

Turner V. and Albassam M.A., Susceptibility of rats to corneal lesions after injectable anesthesia, *Comparative Medicine*, **55** (2): 175 – 182, 2005.

Upton P.K. and Morgan D.J., The effect of sampling technique on some blood parameters in the rat, *Lab Anim*, **9** (2): 85 –91, 1975.

Warren R., *Small Animal Anesthesia*, The C.V. Mosby Company, St. Louis, 1983.

Waynforth H. and Flecknell P, *Experimental and Surgical Technique in the Rat*, Academic Press, New York, 1992.

Wells T, *The Rat: A Practical Guide*, Dover, New York, 1968.

White P., *et al.*, Ketamine - Its pharmacology and therapeutic uses. *Anesthesiology*, **56**: 119 – 136, 1982.

Wiersma J. and Kastelijn J., A chronic technique for high frequency blood sampling/transfusion in the freely behaving rat which does not affect prolactin and corticosterone secretion, *J Endocr*, **107**: 285 – 292, 1985.

Williams L. *et al.*, Continuous infusion of nerve growth factor prevents basal forebrain neuronal death after fimbria fornix transection, *Proceedings of the National Academy of Science*, **83**: 9231 – 9235, 1986.

Wintobe M.W., *et al.*, *Clinical Hematology*, 8th Edition, Lea and Febiger, Philadelphia, 1981.

BIOGRAPHY

With a degree in Veterinary Science from the University of Massachusetts at Amherst, Ms. Heiser has spent over 20 years perfecting these surgical techniques in various academic and industrial settings. Her experience covers many disciplines including toxicology, neuro- & ocular pharmacology, cell biology, and pharmacokinetics. She has authored & co-authored over 15 scientific publications. In addition, she has served over a decade as a member and co-chair of IACUCs. She was appointed and serves as a board member of the New England Branch of AALAS. She is employed as a Lab Animal Resources manager at a major pharmaceutical company and lives in Massachusetts with her husband, two children, dog, rabbit, & bearded-dragon.

Index

flush, 17, 28, 48, 59,68, 69
heparin-coated, 26
hybrid, 27
insertion, 31, 34, 46, 47, 56, 61, 63, 64, 71, 73
leakage, 11
location, 31
perflourocarbon, 28
polyethylene, 26
polyvinyl, 28
pre-constructed, 28
preparation, 17, 28
silicon (silastic®), 27
Tygon®, 28
Caudal, 82, 92
CBV, 66, 94
Cerebral blood flow, 10
Cholinergic, 66, 92
Cidex, 17
Circle of Willis, 32
Clidox, 17
Clots, 26, 27, 48, 68, 69, 93
CNS, 9, 14, 65, 73
Contamination, 15, 26
Continuous infusion, 27, 70–72
Cornea, 92
corneal injury, 18
protection, 18
reflex, 13

Corticosterones, 1
Cranial, 82, 92
Cryosurgery, 24, 92
CYP inhibition, 11

D
Decapicone, 38, 85, 92
Diarrhea, 3, 92
Dilate, 92
Discharge, 3
Disinfectant, 15, 16, 17, 19, 90
cidex, 17
clidox, 17
novalsan, 17
sonacide, 17
sporocidin, 17
Dissection, 82
scope, 20, 21, 86
Distal, 31, 78, 82, 93
Diuresis, 9, 93
Dorsal, 82
recumbency, 39, 93
Dose, 70, 83
analgesia, 9
anesthetic, 3, 7, 9, 10, 11
atropine, 5
cumulative, 92
gavage, 70

Drapes, 16, 17, 18, 37, 40
Dyspnea, 93

E
Ear pinna reflex, 13
Edema, 5, 6, 93
Embolism, 71, 93
Enzootic disease, 3, 93
Euthanasia, 16
Extravasation, 27, 93

F
Fasting, 4
Fat, 10, 11, 12, 41, 79
Fill solution, 26, 72
heparinized glucose, 26
heparinized PVPD, 26
heparinized saline, 26, 46, 48, 56
Fluid volume, 24, 25, 66
Flunixamine, 4
Flunixin, 4
Fur matting, 3

G
Gavage, 70, 86, 93

H
Hair removal, 18, 39
Heart rate, 5, 9, 10, 14, 66, 74